This book describes the principles and equations required for evaluating the performance of an aircraft.

After introductory chapters on the atmosphere, basic flight theory and drag, the book goes on to consider in detail the estimation of climbing performance, the relevant characteristics of power plants, take-off and landing performance, range and turning performance. There is then a short account of the use of vectored thrust and a final chapter on transonic and supersonic flight. The emphasis is on deriving and applying simple analytical expressions rather than on computations from extensive numerical data, but all the exemplary calculations refer to current types of aircraft and the few necessary data are provided. The greater part of the book refers to subsonic aircraft flying at subcritical speeds, which can be dealt with using relatively simple equations, but it is shown that some of this simplicity can be carried over to transonic and supersonic speeds.

The book will be of interest to engineers who wish to improve their understanding of aircraft performance and its evaluation.

Cambridge Aerospace Series

1. J. M. Rolfe and K. J. Staples (ed.): *Flight Simulation*
2. P. Berlin: *The Geostationary Applications Satellite*
3. M. J. T. Smith: *Aircraft Noise*
4. N. X. Vinh: *Flight Mechanics of High Performance Aircraft*
5. W. A. Mair and D. L. Birdsall: *Aircraft Performance*

Cambridge Aerospace Series 5

General editors: David L. Birdsall, Peter Barker

Aircraft Performance

Aircraft Performance

W. Austyn Mair

Emeritus Professor of Aeronautical Engineering,
University of Cambridge

David L. Birdsall

Senior Lecturer, Department of Aerospace Engineering,
University of Bristol

Published by the Press Syndicate of the University of Cambridge
The Pitt Building, Trumpington Street, Cambridge CB2 1RP
40 West 20th Street, New York, NY 10011–4211, USA
10 Stamford Road, Oakleigh, Melbourne 3166, Australia

© Cambridge University Press 1992

First published 1992
First paperback edition 1996

Printed in Great Britain at the University Press, Cambridge

British Library cataloguing in publication data
Mair, W. Austyn (William Austyn) *1917–*
Aircraft performance.
1. Aircraft
I. Title II. Birdsall, David L. (David Lynn) *1940–*
629.13252

Library of Congress cataloguing in publication data
Mair, W. Austyn.
Aircraft performance/W. Austyn Mair, David L. Birdsall.
    p.   cm.—(Cambridge aerospace series)
Includes bibliographical references.
1. Airplanes—Performance.   I. Birdsall, David L.   II. Title.
III. Series.
TL671.4.M35 1991   91-9202 CIP
629.13′2—dc20

ISBN 0 521 36264 4 hardback
ISBN 0 521 56836 6 paperback

Contents

Preface		xv
Introduction		xix
1	**Aerodynamic foundations**	1
1.1	The atmosphere	1
	The International Standard Atmosphere	1
	Equations for variation of temperature, pressure and density with height	2
	Velocity of sound	2
	Pressure height	3
1.2	The forces acting on an aircraft	3
	Propulsive thrust	3
	Lift and drag	4
	Dimensionless coefficients	4
	Angle of incidence	4
	Reynolds and Mach numbers	5
1.3	Air speeds	5
	Equivalent air speed	5
	Calibrated air speed	6
	Measurement of air speed and Mach number	6
2	**Basic flight theory**	7
2.1	Steady flight with zero acceleration	8
	Angle of climb	9
2.2	The drag/lift ratio β	10
	C_D and β as functions of C_L	10
	Introduction to vortex drag	12
2.3	Effects of Reynolds number on β	14
	Justification for neglect of effects	15
2.4	Effects of Mach number on β	17
	Range of Mach number for which effects can be neglected	20

viii Contents

2.5	Introduction to climbing performance	20
	Derivation of angle of climb from thrust and drag	21
	The speed V_{ec} for maximum angle of climb	22
	Rate of climb	22
2.6	Upper limits of height	25
	Absolute ceiling	25
	Limitations imposed by maximum usable lift coefficient	25
	Practicable upper limits of height	25
2.7	Further discussion of the speed V_{ec}	26
	Speed stability and instability	27
3	**Drag equations**	**30**
3.1	Components of drag	30
	Datum drag and lift-dependent drag	31
	3.1.1 Vortex drag	31
	3.1.2 Wave drag	33
	3.1.3 Viscous drag	34
	3.1.4 Spillage drag	34
3.2	Equations representing the drag polar	34
	The simple parabolic drag law	36
	3.2.1 Alternative parabolic expressions	39
3.3	Equations based on the simple parabolic drag law	40
	Equations for minimum β and the speed V_e^* at which this is obtained	40
	The speed ratio $v = V_e/V_e^*$	41
	The ratio β/β_m in terms of v	41
	Example 3.1. Rate of climb	42
3.4	Power required to overcome drag	43
	Speed for minimum drag power	43
	Example 3.2. Minimum drag power	44
4	**Climbing performance**	**46**
4.1	The approximation $\cos \gamma = 1$	47
	Errors in angle and rate of climb due to assuming that $\cos \gamma = 1$	47
	Example 4.1. Effects of assuming that $\cos \gamma = 1$	49
4.2	Climb of aircraft with thrust independent of speed	51
	Speed for maximum rate of climb	51
	Example 4.2. Maximum rate of climb	55
4.3	Climb of aircraft with thrust power independent of speed	55
	Speed for maximum rate of climb	56
4.4	Climb of aircraft with thrust power increasing with speed	58
	Speed for maximum rate of climb	58

4.5	Energy equations	60
	Specific energy	60
	Energy height	60
	Specific excess power	61
4.6	Accelerated climbs	62
	Acceleration factor for correcting rate of climb	62
	Climb at constant EAS or Mach number	62
	Example 4.3. Effect of acceleration on climb performance	64
	Speed for maximum rate of climb, with acceleration	65
4.7	Climb performance in terms of energy height	66
	Gain of energy height in minimum time, or with minimum use of fuel	68
4.8	Maximum angle of climb	69
	Effect of acceleration	71
4.9	Rate of climb in a non-standard atmosphere	72
5	**Power plants**	**74**
5.1	Efficiency of thrust generation	75
	Principle of operation of an air-breathing power plant	75
	Ideal power, propulsive efficiency	76
	Specific thrust	76
	Specific fuel consumption	77
5.2	Turbojet and turbofan engines	77
	Essential features	77
	By-pass flow and by-pass ratio	78
	Gain of efficiency and noise reduction due to by-pass	79
	Gross and net thrust	80
	Reheat or afterburning	80
	5.2.1 Non-dimensional relations	80
	Typical plots of engine characteristics	81
	5.2.2 Maximum thrust of civil turbofans	82
	Variation with speed during take-off	83
	Power law for variation with speed in climb	85
	Power law for effect of varying height	87
	5.2.3 Fuel consumption of civil turbofans	87
	Power law for variation of sfc with speed and height	88
	Effect of thrust reduction below rated value	89
	5.2.4 Military turbofans and propulsion for supersonic civil aircraft	90
	Military turbofans:	90
	Variation of thrust with speed and height	91
	Variation of sfc with speed	93
	sfc at reduced thrust	93

	Effects of reheat	93
	Characteristics of the Olympus 593 turbojet in Concorde	96
5.3	Propellers	97
	Dimensionless coefficients	97
	Advance ratio	97
	Efficiency	97
	Disc loading and propulsive efficiency	98
	Advantages of controllable pitch	99
	Installation effects	99
	Noise and loss of efficiency at high blade tip Mach numbers	101
	Limitation of cruising speed	101
5.4	Turboprops	102
	Shaft power and equivalent shaft power	102
	Use of controllable pitch	103
5.4.1	Maximum shaft power	103
	Power law to represent increase of power with flight speed	103
	Reduction of power with increasing height	104
5.4.2	Fuel consumption	106
	Definition of sfc in terms of equivalent shaft power	106
	Power law for variation of sfc with flight speed	106
	Effect of power reduction below rated value	107
5.5	Propfans and other open-rotor power plants	108
	Advantages of open rotors	108
	Sweptback blades	109
5.5.1	Maximum thrust	112
	Variation with speed during take-off	112
	Power law for variation with speed in cruise or climb	112
	Power law for effect of varying height	112
5.5.2	Fuel consumption	114
	Power law for variation of sfc with speed	114
	Effect of varying height	114
	Effect of thrust reduction below rated value	115
5.6	Piston engines	115
	Supercharging	116
	Variation of power with height	116
5.7	Summary of conclusions	116
5.7.1	Maximum thrust of turbojets, turbofans and propfans	117
5.7.2	sfc of turbojets, turbofans and propfans	117
5.7.3	Turboprops	118

6	**Take-off and landing performance**	**119**
6.1	High-lift devices	120
	Slats and flaps	120
	Effects on lift and drag	120
6.2	Drag of the undercarriage	124
	Empirical data	124
6.3	Effects of ground proximity on lift and drag	125
	Methods of calculation and survey of effects	125
6.4	Drag equations for take-off and landing	126
	Values of K_1 and K_2 in simple parabolic drag law	126
6.5	Take-off procedure and reference speeds	127
	Reference speeds and airworthiness requirements	128
	Summary of required relations between reference speeds	130
6.6	The balanced field length and the take-off transition	130
	Choice of action after engine failure	130
	Take-off after an engine failure	131
	Required climb gradients	132
6.7	The take-off ground run	133
	Simple approximation	133
	Rolling resistance	133
	Equation of motion	134
	Calculation of distance using mean acceleration	134
	Integration of equation of motion	135
	Optimum lift coefficient during ground run	137
	Estimation of ground run when an engine fails	137
	Example 6.1. Take-off ground run	139
6.8	Lift-off, transition and climb	141
	Estimation of airborne part of take-off distance	141
	Example 6.2. Airborne part of take-off distance	144
6.9	Landing procedure	147
6.10	The landing approach	147
6.11	The landing flare	148
	Difficulty of controlling the flare accurately	148
	Estimation of airborne distance	149
6.12	The landing ground run	151
	Braking limits	152
	Estimation of braking distance	153
6.13	Discontinued approaches and baulked landings	154
6.14	The accelerate–stop distance and the balanced field length	155
6.15	Effects of varying air temperature and pressure	157
	WAT curves	157
	6.15.1 Engine characteristics	157
	6.15.2 Take-off distance	158
	6.15.3 Landing distance	158

6.16		Effects of wind	159
	6.16.1	Take-off	159
	6.16.2	Landing	160
7		**Fuel consumption, range and endurance**	**162**
7.1		The phases of a flight	163
7.2		Fuel reserve and allowances	163
7.3		Work done for a specified range	164
7.4		Basic equations for cruise range	166
		Specific range	166
		Total cruise range	166
		Breguet range equation	167
		Example 7.1. Breguet range of turbofan aircraft	168
		Equations for turboprop aircraft	168
7.5		Conditions for maximum cruise range – turbofans	169
	7.5.1	Constant true air speed	170
		The cruise-climb	170
		Example 7.2. Effect of climb angle in cruise-climb	170
		Effects of varying temperature with height	171
	7.5.2	Constant Mach number	172
	7.5.3	Thrust adjustments in a cruise-climb	172
	7.5.4	Speed and height limited by available thrust	173
	7.5.5	Constant height	177
		Alternative cruise procedures	178
7.6		Conditions for maximum range – propellers	181
		The optimum cruise-climb	182
		Constant height	182
7.7		Practical cruise procedures	185
		Choice of Mach number and lift coefficient	185
		Stepped cruise as alternative to cruise-climb	186
		Turboprop aircraft	186
7.8		Calculation of cruise range	188
		Example 7.3. Range of turbofan aircraft, with alternative cruise procedures	189
		Example 7.4. Range of turboprop aircraft, with alternative cruise procedures	192
7.9		Endurance	194
		Specific endurance	194
	7.9.1	Turbofans	195
		Example 7.5. Endurance of turbofan aircraft, with alternative flight procedures	196
	7.9.2	Turboprops	197
		Example 7.6. Endurance of turboprop aircraft, with alternative flight procedures	199

7.10	Effects of climb and descent	200
	Lost range, lost time and lost fuel	200
	Effects on endurance	204
7.11	Effects of engine failure	204
	Reduction of height after engine failure	205
	Example 7.7. Reduction of range of turbofan aircraft caused by engine failure	205
7.12	Effects of wind	207
	Variation of optimum flight speed with wind speed	208
7.13	Variation of payload with range	211
	The payload–range diagram	212

8 Turning performance — 214

8.1	Curved flight in a vertical plane	214
	The load factor n	215
	Relation between acceleration and load factor	215
8.2	Equations for a banked turn	217
8.3	Structural and human limitations on the load factor	218
8.4	Turning limitations due to stalling or buffeting	219
	Minimum radius of turn	221
	Maximum load factor and rate of turn	224
	Example 8.1. Maximum load factor, limited by stalling or buffeting	226
8.5	Rate of climb in a banked turn	228
	Reduction of specific excess power due to increased drag	228
	Example 8.2. Deceleration in a turn due to thrust deficit	231
8.6	The thrust boundary for a banked turn at constant height	233
	Thrust required for a turn without loss of height or speed	233
	Example 8.3. Maximum values of angle of bank and rate of turn, for no loss of speed or height	236

9 Vectored thrust — 237

9.1	Equations for steady flight	239
9.2	Optimum values of θ_F for cruise and climb	241
9.3	Level flight at low speed	241
	Equations for partially jet-borne flight	242
	Optimum jet deflection angle and required thrust	243
	Example 9.1. Thrust conditions required for specified reduction of minimum speed	244
9.4	Vertical take-off and landing	244
9.5	Short take-off	245
	The 'ski-jump' ramp	245
	Calculation of take-off distance from a flat runway	246

xiv   *Contents*

9.6	The use of vectored thrust in a turn	248
	Reduction of minimum radius of turn	248
	Conditions for turn without loss of height or speed	249
9.7	Other uses of vectored thrust in combat	251
10	**Transonic and supersonic flight**	**252**
10.1	Drag	253
	Wave drag	253
	Validity of simple parabolic drag law, with coefficients dependent on Mach number	253
	Shock-induced separation drag at transonic speeds	253
	Drag and drag/lift ratio for a subsonic aircraft at cruising speeds	254
	Drag and drag/lift ratio for supersonic aircraft	257
10.2	Range at high subsonic speeds	263
	Conditions for maximum specific range	265
10.3	Climb and acceleration in supersonic flight	268
	Example 10.1. Supersonic civil aircraft: maximum values of rate of climb and acceleration	271
	Example 10.2. Supersonic combat aircraft: maximum values of rate of climb and acceleration	272
10.4	Range at supersonic speeds	273
	Conditions for maximum range	273
	Example 10.3. Range of supersonic civil aircraft	276
10.5	Turning in supersonic flight	277
	Limitations imposed by maximum usable lift coefficient	277
	Thrust required for turn without loss of height or speed	279
	Example 10.4. Thrust required for turn of supersonic combat aircraft	281

Appendixes

1	List of symbols	283
2	The International Standard Atmosphere	288
3	Conversion factors	291

References   292

Index   296

Preface

The estimation of the performance of an aircraft requires calculations of quantities such as rate of climb, maximum speed, distance travelled while burning a given mass of fuel and length of runway required for take-off or landing. The aim of this book is to explain the principles governing the relations between quantities of this kind and the properties of the aircraft and its power plant. Thus the emphasis is on the development of simple analytical expressions which depend only on the basic aircraft properties such as mass, lift and drag coefficients and engine thrust characteristics. Although extensive numerical data are required for the most accurate estimates of performance in the later stages of a design, the use of such data is not considered here and the data required for use in the simple expressions to be derived are of the kind that would be readily available at the preliminary design stage of an aircraft. Only fixed wing aircraft are considered and the measurement of performance in flight is not discussed.

One of the authors (WAM) has given for many years a short course of lectures on aircraft performance to engineering students at the University of Cambridge. Experience with these lectures has drawn attention to the shortcomings of existing books and to the need for a new book with the aim stated above. The book follows the same approach as the lectures, although it covers a greater range of topics and these are examined in much greater detail. Little previous knowledge of aircraft is assumed and the level of mathematics required should be well within the capabilities of engineering students, even in their first year.

The book should be useful to undergraduate and graduate students of aeronautical engineering and also to engineers working in industry who wish to improve their understanding of the fundamental principles of aircraft performance. Indeed if appropriate data are available for engine thrust and for lift and drag, perhaps from wind tunnel tests, the methods presented here will be found to provide quite adequate accuracy for many industrial requirements.

It is clear that the performance of an aircraft depends on its design and it follows that performance may often be improved by changes in design. This is a subject of great importance but the design of an aircraft involves an integration of several different technologies, aerodynamics, structures, materials, power plants, systems etc. and a proper consideration of any one of these in sufficient depth to reveal its significance in relation to performance would be far beyond the scope of this book. Thus the book is not a textbook on aircraft design, although the use of the overall characteristics mentioned earlier allows some useful inferences to be drawn as to design changes that might be beneficial.

Of the ten chapters in the book, nine are concerned exclusively with flight at subcritical speeds, i.e. speeds that are not so high that there are important effects of increasing Mach number due to compressibility of the air. There are two main reasons for limiting the subject in this way. One is that all civil aircraft in service at present, except Concorde, fly at such speeds; the other reason is that performance at these subcritical speeds can be analysed in a particularly simple way and this forms the main theme of the book. It is shown in Chapter 10, however, that some of the simplicity can be carried over to transonic and supersonic speeds.

In order to simplify the equations and to avoid the need for conversion factors the units used throughout the book are SI, even though international agreements specify that in aircraft *operations* all speeds and heights are measured in knots and feet respectively.

The selection of a set of symbols which does not require some dual definitions, or which satisfies traditions which differ internationally, is an impossible task. The symbols used here have been chosen to agree as closely as possible with those used in the data sheets published by ESDU International plc, with the addition of some further symbols not used by ESDU. Some dual definitions have been unavoidable, but these should not lead to any ambiguities in the text.

It will be noticed that considerable use has been made of the ESDU data sheets and that the reader is encouraged to pursue this major source of guidance, which is based on long experience of many different types of aircraft. The items are continually being improved and updated and indeed it is known that several of the items given here as references are to be replaced in the near future, with new titles and numbers.

The examples given in the book have been chosen to illustrate methods of calculation from the equations, to indicate the magnitude of errors introduced by various approximations and to engender an appreciation of typical values for numerical quantities. Most of the examples refer to transport aircraft of medium or large size, powered

Preface

by turbofans or turboprops, but some examples related to combat aircraft are also included. For some other classes, such as light aircraft, the numerical values would be different but of course the physical quantities which are necessary for the calculations may be expected to be the same for all classes of aircraft.

The authors acknowledge with gratitude the help they have received from many individuals and organisations in the preparation of this book. Particular thanks are due to British Aerospace plc, Dowty Rotol plc and Rolls–Royce plc for the provision of performance data and valuable advice and to Dornier GmbH for data on the Dornier 228. The authors also wish to thank the following organisations for the provision of photographs: Bell Helicopter Textron Inc. for Figure 9.2; British Aerospace plc for Figures 3.4(*b*), 4.4, 8.1, 9.1 and 9.6; British Airways and Adrian Meredith Photography for Figures 6.2, 6.3 and 10.5; GE Aircraft Engines for Figure 5.24; Gulfstream Aerospace Corporation for Figure 3.4(*a*); Lockheed Aeronautical Systems Company for Figure 2.11; NASA Lewis Research Center for Figure 5.23; Rolls–Royce plc for Figure 5.4 and the Saab Aircraft Division of SAAB–SCANIA AB for Figure 7.5.

Finally, the authors wish to express their sincere thanks to Miss Jess Holmes for typing the text on a word processor and to Miss Pamela Eveleigh and Miss Carrie Pharoah for preparing the diagrams for printing.

Introduction

The performance of an aircraft is essentially a statement of its capabilities and a different selection of these will normally be specified for the various categories such as transport, military and light aircraft, even though several common performance factors will feature in every such selection. For the engineer involved in the creation of a new design, these performance features serve as design criteria or at least desirable objectives, whereas late in the design and development stages the sales staff will quote the performance features as the basis for the commercial strength of the emerging aircraft. For either reason the performance will be stated in terms of quantities such as direct operating cost (DOC), maximum range for various payloads and fuel loads, cruising speed and airport requirements for landing and take-off. While the sales and design attitudes will be distinct, although related, this book addresses the early stages of the design process which must also bear considerable allegiance to performance as viewed by a potential customer.

The estimation of performance proceeds in stages, starting with parametric studies based on simple assumptions and progressing to more refined calculations as the main features of the design become established and the confidence in data grows. Estimation techniques are important not only because they allow the engineering team to proceed while data are crude or speculative, but also because construction of the new aeroplane will begin well in advance of the engineering refinements, and if there is accuracy in the early estimations this will be rewarded by a reduction in modifications as the fabrication and assembly effort progresses toward regular production. The reliability of these early predictions will also serve to identify lines of necessary research and development when the achievable performance appears to depart from that which is desired.

The aim of this book is to set out the basic principles of performance estimation and to develop methods of analysis and calculation that are based on simple assumptions of the kind used in the early stages of performance estimation for a new aircraft. Unless otherwise stated it will be assumed that the aircraft is flying in still air. The wind velocity relative to the ground becomes important for performance only during take-off or landing and in considering either the range of an aircraft or the flight conditions required to give maximum range. These effects of wind are considered in Chapters 6 and 7.

1
Aerodynamic foundations

The study of performance of aircraft requires a careful pursuit of accurate aerodynamics because, with the exception of braking, propulsive and gravitational forces, all forces which aid or inhibit flight are aerodynamic in origin. If these aerodynamic forces are to be described with accuracy it will first be necessary to describe appropriate features of the atmosphere and then to define in the simplest terms the forces which can be produced on an aircraft as it penetrates the air at speed. This chapter is devoted to these needs, standard properties of the atmosphere are developed in mathematical terms and definitions are provided for aerodynamic forces and flight speeds.

1.1   The atmosphere
In an atmosphere that is either stationary or in uniform motion gravity causes a reduction of air pressure p with increasing height h given by

$$\mathrm{d}p/\mathrm{d}h = -g\rho \tag{1.1}$$

where g is the acceleration due to gravity and ρ is the air density, which also varies with height. The variation of g with height, i.e. with distance from the centre of the earth, will be neglected here, although it should be recognised that in extreme cases such as the very high altitude high speed flight of an aircraft like Concorde, not only may the variation of g have an appreciable effect but also the upward centrifugal force due to curvature of the earth can amount to a measurable percentage of the total force normal to the flight path. These effects will be discussed further in § 10.4.

Equation (1.1) can be integrated to find the pressure at any height, but only if the density function is accurately known and this implies allowance for a change of temperature with height. Appendix 2 gives details of the *International Standard Atmosphere* (ISA) in which the temperature is assumed to vary linearly with height up to 11 km above sea level and then to remain constant up to 20 km. The equations required for calculating the variation of pressure and density in the ISA, subject to that assumption, are derived below, using suffix 0 to denote conditions at sea level.

In a region where the air temperature T varies linearly with the height h above sea level

$$T = T_0 - Lh, \tag{1.2}$$

where $L = -dT/dh$ is the *lapse rate*. Assuming air to be a perfect gas, the density is given by the equation of state as

$$\rho = p/(RT) = p/[R(T_0 - Lh)], \tag{1.3}$$

where R is the gas constant for air. Equation (1.1) then gives

$$dp/p = -g \, dh/[R(T_0 - Lh)], \tag{1.4}$$

from which can be found the pressure ratio related to sea level

$$p/p_0 = (1 - Lh/T_0)^{g/(RL)} = (T/T_0)^{g/(RL)} \tag{1.5}$$

and the density ratio

$$\rho/\rho_0 = (T/T_0)^{(g/(RL))-1}, \tag{1.6}$$

both of these now being expressed in terms of the temperature ratio and constants which are well known.

In the higher region of the atmosphere, where the air temperature is constant and equal to T_1, the air density is given more simply by

$$\rho = p/(RT_1),$$

while the new pressure function is derived from

$$dp/p = -g \, dh/(RT_1) \tag{1.7}$$

so that

$$p/p_1 = \exp[-g(h - h_1)/(RT_1)] \tag{1.8}$$

where p_1 is the pressure and h_1 is the height at the lower boundary of this region.

The velocity of sound a will also be of interest later when flight at transonic and supersonic speeds is considered, because for an isentropic compression $dp/d\rho = a^2$ and thus $1/a^2$ is a measure of the compressibility of the air. Since $a^2 = \gamma_a RT$, where γ_a is the ratio of specific heat capacities, the ratio of the speed of sound at altitude to that at sea level is

$$a/a_0 = (T/T_0)^{1/2} \tag{1.9}$$

and again this is expressed in terms of the temperature function.

An *altimeter* connected to a static pressure tube is designed to measure the absolute pressure of the undisturbed air at the height at which the aircraft is flying. If the instrument is set to read zero height when the air pressure is equal to the sea-level pressure in the ISA, it will read the true height above sea level only if the atmosphere has the same properties as the ISA. In a non-standard atmosphere the altimeter with its zero set in this way will read what is known as the *pressure height*, namely the height in the ISA at which the pressure would be equal to the current measured pressure. The true height may be different and not available from the simple pressure law which is implicit in the altimeter mechanism. It is worth noting that it is standard practice for pilots to follow a pressure height, knowing that all nearby aircraft (using similar instruments) will be subject to the same height errors and can thus retain safe height separations even though their true heights all differ from those indicated.

The true air pressure at altitude is normally available from sensors along with the pressure height derived from it. Clearly, if the air temperature is measured then both the local air density and the local velocity of sound can also be deduced. It will be assumed throughout this book, unless otherwise stated, that the aircraft is flying in air which has the properties of the ISA.

1.2 The forces acting on an aircraft

The motion of an aircraft in steady flight is determined by

(i) its weight,
(ii) the propulsive thrust exerted on the aircraft by the power plant,
(iii) the aerodynamic force generated on the aircraft by its motion through the air.

It will be assumed here that there is no interaction between (ii) and (iii), so that each can be considered independently. This is a simplification which is useful in developing the basic performance equations but is not strictly correct, because there is an influence of the jet efflux or propeller slipstream on nearby surfaces and also an effect of the engine intake flow on the external flow near the intake. The subject will be discussed further in § 3.1.4.

The velocity of the aircraft relative to the undisturbed air is a vector directed along the instantaneous flight path having a magnitude V which is the *True Air Speed* (TAS). In the usual conditions of steady flight the vector lies in the plane of symmetry of the aircraft, i.e. there is no sideslip. The aerodynamic force vector then also lies in the plane of symmetry and it is usual to resolve this force into two components

Aerodynamic foundations

as shown in Figure 1.1:

(i) The *lift* L has a direction normal to the flight path,
(ii) The *drag* D is directed backward along the flight path.

It is important to note that the lift is not vertical unless the flight path is horizontal.

The lift and drag are commonly known as forces, although strictly they are components of a single force vector. They are usually expressed in terms of dimensionless coefficients, defined as

$$\text{Lift coefficient} = C_\text{L} = L/(\tfrac{1}{2}\rho V^2 S) \tag{1.10}$$

and

$$\text{Drag coefficient} = C_\text{D} = D/(\tfrac{1}{2}\rho V^2 S), \tag{1.11}$$

where S is the *gross wing area*, i.e. the total area of the wing (but not the tailplane), including that part of the wing which may be regarded as 'covered' by the fuselage. If the analysis is confined to steady flight, ignoring the dynamics of manoeuvres and disturbances, then L, D, C_L and C_D must refer to the aircraft in the 'trimmed' state, i.e. with the controls adjusted to give zero moment about the centre of gravity.

Figure 1.1 shows a datum line which is fixed in the aircraft and is inclined upward at an angle α to the flight path. This angle is known as the *angle of incidence* or *angle of attack* and may be defined relative to any specified datum line in the aircraft. Often a longitudinal axis of the fuselage is selected when defining the incidence of a complete aircraft, whereas the mean chord line of a lifting surface is the usual choice when the lift on such a surface is considered in isolation. For an aircraft of given shape, with its scale specified by the single dimension of mean wing chord $\bar{c}$, both L and D are functions of the six variables α, $\bar{c}$, ρ, a, V and μ, where μ is the viscosity of air, sometimes known

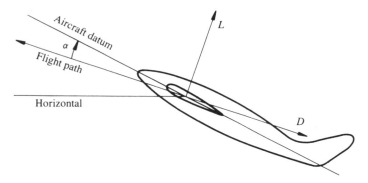

Figure 1.1. Components of aerodynamic force.

as the dynamic viscosity to distinguish it from the kinematic viscosity μ/ρ. Dimensional analysis then shows that both of the dimensionless coefficients C_L and C_D are functions of the three dimensionless parameters

Angle of incidence α
Reynolds number $Re = \rho V \bar{c}/\mu$
Mach number $M = V/a$.

The effects of varying Re on C_L and C_D, due to variations in speed and height, are relatively small within the range of flight speeds and heights to be considered for most aircraft and are often neglected in preliminary estimates of performance. It may also be justifiable, for these early estimates, to assume that the Mach number M is sufficiently low for the effects of its variation to be negligible. Nevertheless the effects of varying Re will be considered further in Chapter 2 and the dependence of C_D on Mach number will be discussed in that chapter and also in Chapter 10.

A further simplification will be made, except in Chapter 9 where vectored thrust is a special consideration, namely that the propulsive thrust F exerted on the aircraft by the power plant acts along the flight path. With the power plant and its thrust line fixed in the aircraft this assumption can be correct at only one value of α, but it will be shown in Chapter 9 that the errors introduced by this simplification are usually small.

1.3   Air speeds

In deriving simple forms of the performance equations there is an obvious appeal in the use of *Equivalent Air Speed* (EAS), denoted by the symbol V_e, because it allows coefficients and many of the equations to be expressed in a form that is independent of height. Its definition rests on the recognition that if the dynamic pressure $\frac{1}{2}\rho V^2$ at altitude is equal to the dynamic pressure at sea level at a speed V_e, the two speeds V and V_e are related by

$$\tfrac{1}{2}\rho V^2 = \tfrac{1}{2}\rho_0 V_e^2,$$

from which is determined

$$V_e = (\rho/\rho_0)^{1/2} V = \sigma^{1/2} V, \tag{1.12}$$

where σ is the relative air density at altitude, the ratio of the density there to the value ρ_0 at sea level in the ISA. Thus in Equations (1.10), (1.11) and many others it is convenient to employ $\rho_0 V_e^2$ instead of ρV^2 and thereby use a standard air density. Reversion to expressions showing True Air Speed at altitude is effected by using Equation (1.12) and in all later chapters the word 'speed' will mean the true air speed V unless otherwise stated.

A measurement of air speed that is closely related to EAS is *Calibrated Air Speed* (CAS) which is the reading seen on a correctly calibrated *Air Speed Indicator* (ASI) connected to pitot and static tubes which are free from error. The ASI measures the difference between the pressure p_t in the pitot tube and the static pressure p of the undisturbed air. For subsonic flight speeds and using $\gamma_a = 1.4$ for air Pankhurst & Holder (1952) show that the equations for isentropic flow lead to the result

$$p_t - p = \tfrac{1}{2}\rho V^2 f(M) \tag{1.13}$$

where

$$f(M) = 1 + M^2/4 + M^4/40 + M^6/1600 + \ldots$$

The implication here is that if the $f(M)$ term in Equation (1.13) were ignored there would be an error in the pressure difference of about 6.4% at $M = 0.5$, for example, and thus an error in the air speed of about 3.1%. For supersonic speeds it is necessary to make allowance for the non-isentropic compression at the shock wave ahead of the pitot tube and the necessary equations for this are given by Liepmann & Roshko (1957).

The Mach number M can be obtained directly from measurements of $(p_t - p)$ and p, without considering the temperature, because for subsonic speeds with $\gamma_a = 1.4$ the equations for isentropic flow lead to the result

$$M^2 = 5[(p_t/p)^{2/7} - 1]. \tag{1.14}$$

It is also useful to note that

$$M^2 = \tfrac{1}{2}\rho V^2/(0.7p). \tag{1.15}$$

An air speed indicator is calibrated so that it displays true air speed correctly in the ISA at sea level and allows for the effects of compressibility at sea level as given by Equation (1.13). At any other height the instrument would display the equivalent air speed V_e if the Mach number were small enough to make $f(M) \approx 1$ in Equation (1.13). In general this condition is not satisfied and moreover, if V_e is fixed the Mach number increases with height, thus also increasing the value of $f(M)$. Then the CAS is greater than the EAS by an amount that increases with Mach number and altitude.

International agreements on Air Traffic Control require the measurement of all operating heights in feet and all speeds in knots. (Note that 1 knot = 1 nautical mile per hour and as 1 nautical mile is 1852 m, 1 knot = 0.5144 m/s.) In order to avoid the intrusion of unnecessary conversion factors these units will not be used here; standard SI units will be used, namely heights in metres or km and speeds in m/s.

2
Basic flight theory

The first chapter has given an introduction to the characteristics of atmospheric air and has provided a valid basis for the expression of the aerodynamic force developed during flight in that air, but there has not yet been any attempt to consider the balance of forces necessary to satisfy the laws of mechanics. Except for Chapters 6 and 8 and parts of Chapters 4 and 10, this book is directed mainly towards flight with zero or negligible acceleration so that the equations to be developed are those of *statics*, not *dynamics*.

Consideration of the effects of varying speed and altitude on the aerodynamic force on an aircraft can be greatly simplified by examining the dependence of the lift and drag coefficients on the Reynolds and Mach numbers. This dependence has already been mentioned briefly and is discussed further in this chapter, where it is shown that for a given aircraft the variations of Reynolds number caused by changing speed and altitude are likely to have only small effects. With increasing Mach number in the high subsonic range there is usually a large increase of drag coefficient and this important effect is introduced briefly, deferring a more detailed account until Chapter 10.

An important measure of the aerodynamic efficiency of an aircraft is the ratio L/D of lift to drag, since there is always a desire to create lift with as little cost in drag as possible. In this chapter the effects of this ratio (or its reciprocal D/L) on some important performance parameters are examined and it is shown that there is a minimum value of D/L which is especially important. The concept of a ceiling of flight is also considered. This is the maximum achievable altitude and it depends not only on the aerodynamic efficiency L/D but also on propulsive thrust at high altitudes, aircraft weight and the limitations imposed by the atmosphere as it becomes rarefied at high altitude.

This book is concerned mainly with speed, range, rate of climb etc. in steady flight and it is implied throughout that the aircraft is stable in the flight regime considered. A general discussion of stability is outside the scope of the book but there is one particular aspect, speed stability, which is closely related to the balance between thrust and drag and is introduced briefly in this chapter.

2.1 Steady flight with zero acceleration

In addition to the need for stability that has already been mentioned, a further requirement for steady flight is that the moment about the centre of gravity of the aircraft (CG) must be zero. This means that the resultant force vector, which is the vector sum of the propulsive thrust and the aerodynamic force, must pass through the CG. In practice this means that the elevator angle must be continually adjusted so that by changing the tailplane lift this condition is satisfied. With a stable aircraft an increase of speed from one steady state to another usually causes a positive (nose-up) pitching moment if there is no movement of the elevator, and in order to counteract this moment and achieve equilibrium in the steady state at the higher speed a downward deflection of the elevator is needed. If the flight path is horizontal the lift remains equal to the weight, the change of tailplane lift being compensated by an adjustment of α to change the wing lift.

As stated in Chapter 1 it will be assumed that the propulsive thrust F acts along the flight path, although this can be strictly correct at only one value of α. With this assumption the line of action of F may still be above or below the CG. The thrust then generates a pitching moment about the CG which is counteracted in steady flight by adjustment of the elevator so that the *aerodynamic* force vector produces a pitching moment of the required sense and magnitude. The aerodynamic force is of course the resultant of all the forces generated by the wing, fuselage, tailplane etc. and for performance calculations it is only this resultant force that needs to be considered, not the individual components. Strictly, in order to justify another assumption given in Chapter 1, that there is no interaction between thrust and drag, it is necessary to assume that the line of action of the thrust F does pass through the CG. If this condition is not satisfied a change of thrust requires a deflection of the elevator, in order to maintain the necessary moment due to the aerodynamic force and in general this causes some change of drag. The effect is usually very small and in practice the position of the thrust line is not important.

Figure 2.1 shows the forces acting on an aircraft which is flying along a path inclined at an angle γ to the horizontal. The aerodynamic force is resolved into two components L and D as in Chapter 1 and the line of action of the thrust F is shown as passing through the CG, where the forces L and D and the weight W intersect. Apart from the point already mentioned about the independence of thrust and drag there is no need to assume that F passes through the CG and the force

Steady flight with zero acceleration

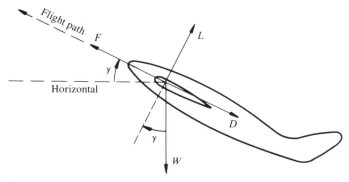

Figure 2.1. Forces on an aircraft in a steady climb.

equations to be derived are unchanged if the line of action of F is displaced.

It is important to distinguish between the angle γ shown in Figure 2.1 and the angle of incidence α shown in Figure 1.1. The angle γ is known as the *angle of climb* or *climb gradient* and is positive in the sense shown, but the equations to be derived are of course valid when γ is negative and the aircraft is descending. This angle γ indicates the direction of motion of the aircraft relative to the horizontal, whereas α relates a datum line that is fixed in the aircraft to the direction of motion. Although γ can be either positive or negative the datum line used for defining α is usually chosen so that α is almost always positive in steady flight. The angle α is not shown in Figure 2.1 but in the case represented there it is quite small.

With a straight flight path and zero acceleration the forces shown in Figure 2.1 are in equilibrium so that

$$L = W \cos \gamma \qquad (2.1)$$

and

$$F = D + W \sin \gamma. \qquad (2.2)$$

A useful dimensionless equation may be derived from these two equations by introducing the ratios

$$\beta = D/L = C_D/C_L \quad \text{and} \quad f = F/W.$$

Then, since $D/W = \beta L/W$ and $L/W = \cos \gamma$,

$$f = D/W + \sin \gamma = \beta \cos \gamma + \sin \gamma. \qquad (2.3)$$

Equation (2.3) is the basic equation for all calculations of performance in steady flight. It relates two important ratios which are often quoted as measures of efficiency or performance capacity, namely lift/drag

10   Basic flight theory

and thrust/weight. Two simple cases may be noted:
(a) level flight ($\gamma = 0$), for which $f = \beta$,
(b) gliding flight ($F = 0$), which requires $\tan \gamma = -\beta$.

Thus in contrast to the conventional use of *angle of climb* for powered aircraft the *angle of glide* (with zero thrust) is used for gliders and is conventionally taken to be positive when descending. It requires only a change of sign for γ and is seen here to be equal to $\tan^{-1}\beta$ (or β, when β is small).

Equations (1.10), (1.12) and (2.1) give the lift coefficient as

$$C_L = (W \cos \gamma)/(\tfrac{1}{2}\rho V^2 S) = (w \cos \gamma)/(\tfrac{1}{2}\rho_0 V_e^2), \qquad (2.4)$$

where $w = W/S$ is introduced as the *wing loading*, a measure of overall pressure on the wing, typically of order $5000\,\text{N/m}^2$ for military and for large civil aircraft when fully loaded, but as low as $1500\,\text{N/m}^2$ for some aircraft designed for short take-off and landing (STOL). It is important to note that the wing loading of a given aircraft is not constant, it decreases as fuel is consumed. For a long-range transport aircraft the weight of fuel used may be as much as 40% of the take-off weight, so that the wing loading may drop from 5500 to $3300\,\text{N/m}^2$ over the course of a long flight.

2.2 The drag/lift ratio β

For given values of the Reynolds and Mach numbers Re and M the lift and drag coefficients C_L and C_D depend only on the angle of incidence α and typical curves showing these relations are given in Figure 2.2, which refers to a civil transport aircraft in the cruising configuration, with a wing of aspect ratio about 9. (Aspect ratio will be defined and explained in § 3.1.1.) It should be noted that the quantities plotted in Figure 2.2 are C_L and $10C_D$, because C_D is of order $C_L/10$. In performance calculations the incidence α is usually of little interest and it can be eliminated by using the data of Figure 2.2 to plot C_D against C_L, as in Figure 2.3. Two particular values of α, $2°$ and $13°$, are shown in both the figures and in Figure 2.3 different scales are again used for C_L and C_D. The curve relating C_D to C_L is known as the *drag polar* for the aircraft because, if the scales of C_D and C_L are the same, any point on the curve gives in polar coordinates the magnitude and direction of the resultant aerodynamic force, expressed as a dimensionless coefficient

$$(\text{Force})/(\tfrac{1}{2}\rho_0 V_e^2 S).$$

Thus the straight line drawn from the origin to any point on the curve represents this non-dimensional force vector and the slope of the straight line is equal to the ratio $\beta = C_D/C_L$, the reciprocal of the ratio L/D which is a measure of aerodynamic efficiency.

The drag/lift ratio β

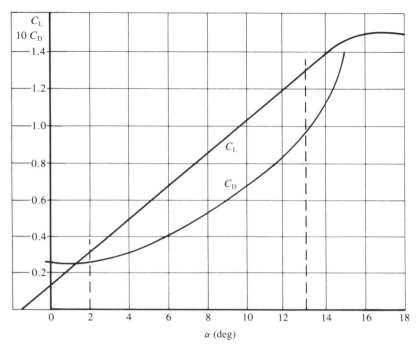

Figure 2.2. Typical lift and drag coefficients for an aircraft.

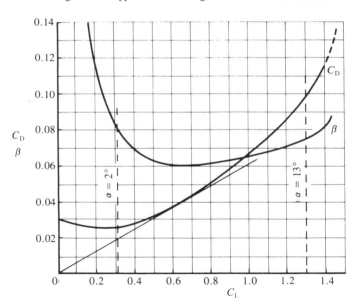

Figure 2.3. Typical variation of C_D and β with C_L.

Basic flight theory

Figure 2.3 also shows the drag/lift ratio β plotted against C_L and an important feature of this curve is the minimum value of β, occurring in this case at $C_L \approx 0.67$. The figure also shows that the tangent from the origin to the drag polar touches the curve at this value of C_L, the slope of the tangent being equal to the minimum value of β, in this case 0.06.

The shape of the curve shown for β in Figure 2.3 deserves some scrutiny and reference to Figure 2.2 is necessary for an understanding of that shape. The lift coefficient C_L is zero at a slightly negative incidence and increases linearly with α until the stall starts to develop, in this case at $\alpha \approx 14°$. The drag coefficient C_D is minimum at a small positive incidence and of course never falls to zero. The incidence is defined in terms of an arbitrarily defined datum line in the aircraft, as shown in Figure 1.1, and the values of α for $C_L = 0$ and for minimum C_D obviously depend on that definition. With any definition commonly used C_L is usually zero for a negative value of α but C_D may sometimes be minimum for a value of α that is close to zero or even negative. In any case, as $C_L \to 0$ it is clear that $\beta \to \infty$ and this conclusion is independent of the choice of datum used for defining α.

Figure 2.2 shows that as α rises above about 14° the slope of the C_L curve falls to zero and becomes negative. This is an indication that separation of the boundary layer has occurred on the upper surface of the wing, i.e. the wing has stalled. In other cases, with different wing designs, the stall may occur at a higher or lower incidence and C_L may either fall sharply or continue to rise at a much reduced rate. The stall also causes a rapid increase of C_D which continues up to much higher values of α, as indicated by the part of the drag polar shown as a broken line in Figure 2.3. The stalled regime is outside the range of normal flight conditions and is not normally considered in performance calculations, although for some military aircraft there is a requirement for operation in a partially stalled state. For a discussion of stalling, reference may be made to Chapter 4 of Küchemann (1978) and Chapter 5 of Thwaites (1960).

Apart from stalling, which occurs only at the top end of the C_L range, there is one factor which has a dominant influence on the shape of the β curve at the higher values of C_L. The next chapter which deals with components of drag will show in greater detail that an important contribution comes from the formation of a vortex sheet trailing behind the wing and a more complete account has been given by Clancy (1975). For a given flight speed the strength of the vortex sheet is proportional to the lift developed by the wing and so is a *consequence* of the lift, but it also *influences* the development of lift at the wing because the vorticity induces a velocity field at the wing that is vertically downward (when the flight path is horizontal) and is called

The drag/lift ratio β

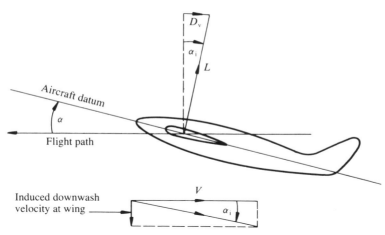

Figure 2.4. Drag due to trailing vortex sheet.

the *downwash*. Thus the lift and trailing vorticity are mutually dependent and, because of the downwash, the local velocity vector at the wing is tilted through an angle α_i called the *induced incidence* as shown in Figure 2.4. The lift vector is normal to the local velocity vector and is tilted back by the angle α_i, giving a force component $D_v = L\alpha_i$ which is parallel to the true flight direction and is thus a drag force. The contribution of this force to the drag coefficient is the *vortex drag coefficient* $C_{DV} = C_L \alpha_i$ and since for a given flight speed

$$\alpha_i \propto L \propto C_L,$$

it must follow that

$$C_{DV} \propto C_L^2.$$

At high lift coefficients C_{DV} makes a major contribution to the total drag coefficient C_D and this is the main reason for the existence of a minimum in the β curve in Figure 2.3 and for the following features of the C_D and β curves at high C_L, but below the stalling incidence:

(a) the positive slope and pronounced upward curvature of the C_D curve in Figures 2.2 and 2.3,
(b) the relatively low curvature of the β curve in Figure 2.3 (because $\beta = C_D/C_L$ and $C_{DV}/C_L \propto C_L$).

Equation (2.3) relates the angle of climb γ to two important ratios, $f = F/W$ and $\beta = D/L$, and in order to make further progress in calculating performance it is necessary to understand how these ratios depend on the flight conditions. The propulsive thrust F can be adjusted by the pilot up to the maximum available and the dependence of this maximum on aircraft speed and height will be discussed in general terms in Chapter 5. More detailed information for a particular power plant is usually obtainable either from engine design calculations or from measurements on a test bed.

In considering the effects of speed and height on the ratio β the first point to note is that C_L and C_D depend not only on the incidence α but also on the Reynolds and Mach numbers Re and M, as explained in Chapter 1. Thus all the curves shown in Figures 2.2 and 2.3 depend in general on Re and M. For given values of w, $\cos \gamma$ and the equivalent air speed V_e and for any specified values of Re and M for which β is known as a function of C_L, Equation (2.4) can be used to find β. For a given aircraft the values of Re and M for any given V_e depend only on height, so that for any given height and for given values of w and $\cos \gamma$, the ratio β can be plotted against V_e. A simplification can be introduced if γ can be assumed to be small, so that $\cos \gamma \approx 1$. The curve relating β and V_e will then apply to all values of γ that are not so large as to make the assumption invalid. The term $\cos \gamma$ appears in both of the Equations (2.3) and (2.4) and it will be shown in Chapter 4 that in consequence the error in the calculated angle of climb due to the assumption that $\cos \gamma = 1$ is usually only about 1° or 2° even when γ is as large as 45°. The assumption will be made in the remainder of the book, unless otherwise stated, noting that if the value of γ calculated for $\cos \gamma = 1$ is regarded as a first approximation the exact value can easily be found by iteration.

A further simplification can be made if it can be assumed, at least in the early stages of estimating performance, that the effects of varying Re and M can be neglected, for then a single curve relating β to V_e is valid at all heights. A closer inspection of these effects, especially on C_D, will now be made.

2.3 Effects of Reynolds number on β

The definition of Reynolds number given in Chapter 1 shows that for any given aircraft

$$Re \propto \rho V/\mu \propto \sigma^{1/2} V_e / \mu, \tag{2.5}$$

in which the viscosity μ is independent of pressure and depends only on temperature. The values of μ given for the ISA in ESDU 77021 (1986) are calculated from the empirical expression

$$\mu = (1.458 \times 10^{-6}) T^{3/2} / (T + 110.4) \text{ N s/m}^2, \tag{2.6}$$

where T is the air temperature in kelvin units and within the range of temperatures to be considered for the ISA the Equation (2.6) is found to be closely equivalent to a simple power law

$$\mu \propto T^{0.8}.$$

Effects of Reynolds number on β

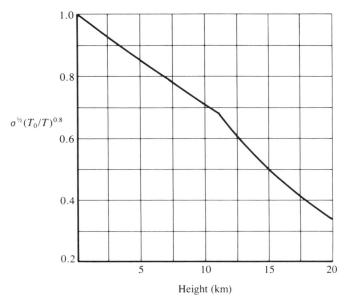

Figure 2.5. Variation of Reynolds number function with height.

Hence the relation (2.5) may be replaced by

$$Re \propto \sigma^{1/2}(T_0/T)^{0.8}V_e,$$

which allows relative values of Re to be determined by looking at the variation of $\sigma^{1/2}(T_0/T)^{0.8}$ with height in the ISA as shown in Figure 2.5. Here it is seen, for example, that for any given V_e the ratio of the Reynolds numbers at heights of 12 and 2 km is

$$Re_{12}/Re_2 = 0.63/0.94 = 0.67. \qquad (2.7)$$

In order to obtain an indication of the effect of this change of Reynolds number on C_D and hence on β, a long-range civil transport aircraft will be considered as an example. For such an aircraft in the cruising condition Bowes (1974) has given a breakdown of the components of drag, showing that roughly 50% of the total drag is due to skin friction. This part of the drag is approximately proportional to the drag of a flat plate of equivalent area and so varies with Reynolds number in the same way as the drag of the flat plate. For Reynolds numbers between 10^6 and 10^8 Duncan, Thom & Young (1970) have shown that the drag coefficient of a flat plate with fully turbulent boundary layers is very nearly proportional to $Re^{-1/6}$. The other components of the drag coefficient of the aircraft, apart from skin friction, are nearly independent of Reynolds number so that for any given change of Reynolds number the percentage change of C_D will be about half the percentage change of flat plate drag coefficient.

16   Basic flight theory

The long-range transport aircraft that is being considered might be expected to cruise at a height in the region of 10–12 km. For reasons of economy which will be discussed in Chapter 7 it would not cruise at the much lower height of 2 km, but as an illustration of the effects of varying Reynolds number the values of C_D for the aircraft will be compared over the extreme height range from 2 to 12 km. Using Equation (2.7) the ratio of drag coefficients for the flat plate is

$$C_{D2}/C_{D12} = (Re_{12}/Re_2)^{1/6} = 0.936 = 1 - 0.064,$$

i.e. there is a reduction of about 6%, so that C_D for the aircraft at any given V_e would be reduced by only about 3% if the height were reduced from 12 to 2 km. The skin friction drag of an aircraft is not sensitive to incidence or lift coefficient, so that the reduction of C_D found here can be regarded as being at constant C_L and the percentage reduction of β would be the same. Since for cruising flight the only height changes that are of practical interest are much smaller than this, the result indicates that effects of varying Reynolds number on the curve relating β to V_e can be neglected in calculations of cruising performance, provided that the values of β are estimated for an appropriate mean value of the Reynolds number. For civil aircraft in climb or descent and for military aircraft which cannot keep to a narrow efficient range of heights, the changes of Reynolds number may have more significant effects and for these cases the flight path should be considered in segments if high accuracy is required in performance estimates, and an appropriate mean Reynolds number should be selected for each segment. For convenience in calculating Re, reference may be made to ESDU 68046 (1977) which gives Reynolds number per metre at $M = 1$ for a range of heights. For any given height Re is of course proportional to M.

In this discussion of the effects of varying Reynolds number it has been assumed that the boundary layers on the aircraft are almost entirely turbulent, so that the skin-friction drag coefficient is nearly proportional to that of a flat plate with fully turbulent boundary layers. This assumption is correct for aircraft of the present day (1990) but in the future it may be possible to maintain substantial areas of laminar flow, at least on the wing, giving a very useful reduction of drag. The sensitivity of C_D to a change of Reynolds number would then be increased because the skin-friction drag coefficient of a flat plate with entirely laminar boundary layers is proportional to $Re^{-1/2}$, in contrast to the $\frac{1}{6}$th power law mentioned earlier for turbulent boundary layers. Moreover the position of transition from laminar to turbulent flow on a wing depends both on lift coefficient and on Reynolds number, so that the estimation of the effect on C_D (and thus on β) of varying Re is much more complex when substantial areas of laminar flow may be expected.

Effects of Mach number on β

In the remainder of the book it will be assumed that any effects of Reynolds number on the curve relating β to V_e can be ignored, because the aircraft boundary layers are mainly turbulent and the curve is estimated for an appropriate mean value of the Reynolds number for the flight condition being considered.

2.4 Effects of Mach number on β

In order to estimate the effects of varying Mach number on the relation between β and V_e it should be noted first that

$$M = V/a = [V_e/a_0][(T_0/T)(\rho_0/\rho)]^{1/2} = (V_e/a_0)(p_0/p)^{1/2}, \qquad (2.8)$$

where a is the velocity of sound, p is the atmospheric pressure and suffix 0 refers to sea level. Clearly, height is important and in Figure 2.6 lines of constant Mach number are drawn showing, for example, that if $V_e = 200$ m/s the Mach number increases from 0.6 near sea level to 0.8 at a height of 5 km and 1.0 at 8 km.

As the Mach number increases above about 0.8 the drag coefficient of an aircraft increases substantially, because shock waves begin to form on the wing and as these become stronger with further increase in Mach number they cause separation of the boundary layer. The value of M at which the steep drag rise starts is often known as the drag divergence Mach number M_D and this usually rises with either increasing sweepback angle of the wing or decreasing thickness/chord

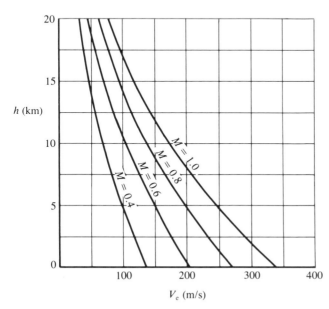

Figure 2.6. Lines of constant Mach number.

ratio. In general, M_D also decreases to some extent as C_L increases, but there is often a range of low lift coefficients for which it is nearly constant. Figure 2.7 shows some results of wind tunnel measurements given by Lock (1986) for a wing of modern design mounted on a body. The wing had a sweepback angle of 28° at the leading edge and 14° at the trailing edge and the quantity ΔC_D that is plotted is the increase of drag coefficient above the value for $M = 0.7$. (This value of M is used as a datum because for lower values there is no appreciable variation of C_D with M at constant C_L.) Lock gives results for lift coefficients of 0.2, 0.3, 0.4, 0.5 and higher values and these show that for any given Mach number the differences of ΔC_D between lift coefficients of 0.2, 0.3 and 0.4 are within the experimental scatter, whereas for values of C_L above about 0.4 the increment in C_D becomes more dependent on C_L as illustrated by the single curve for $C_L = 0.5$. The results are given only up to $M = 0.82$ but for the purpose of illustrating the effect on the curve relating β to V_e the curve for the lower values of C_L in Figure 2.7 has been extrapolated up to $M = 0.85$ as shown by the broken line.

In Figure 2.8 the curve shown as a full line gives the relation between β and V_e as calculated from Figure 2.3 for a wing loading of 5000 N/m². This curve has been labelled ($M = 0$) because the curves given in Figure 2.3 make no allowance for any increase of C_D at high

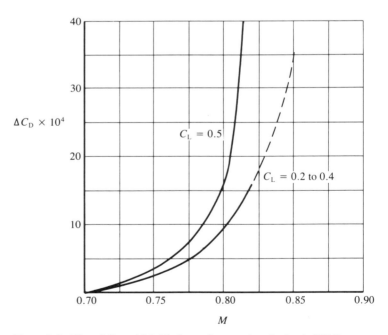

Figure 2.7. Rise of C_D at high Mach number, as given by Lock (1986).

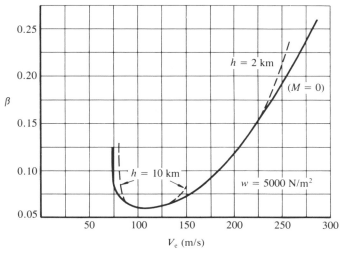

Figure 2.8. Effect of Mach number on β–V_e relation.

Mach numbers. The two curves shown as broken lines in Figure 2.8 give values of β at heights of 2 and 10 km as calculated from Figures 2.3 and 2.7, making the assumption that the curve shown in Figure 2.7 for lift coefficients from 0.2 to 0.4 is also valid when C_L is as low as 0.125 (corresponding to $M = 0.85$ at a height of 2 km).

A further effect that needs to be considered is the loss of maximum lift coefficient C_{Lmax} with increasing Mach number in a range of M that extends down to about 0.2, as this can have a marked influence on β at low values of V_e. This occurs because the peak local relative air velocity over the upper surface of a wing at a high lift coefficient is much greater than the flight speed, so that the local velocity can approach or exceed the velocity of sound for quite low values of the flight Mach number. The effect on complete aircraft has been explored mainly for take-off or landing configurations with flaps and slats extended, e.g. by Fiddes, Kirby, Woodward & Peckham (1985), but earlier measurements on simple aerofoils without flaps have been reported by Hoerner & Borst (1975), and these confirm the well known fact that critical Mach number and these local effects on C_{Lmax} are strongly dependent on the shape of the particular aerofoil section employed. Clearly it is important to obtain accurate data on the effects of Mach number on relations of the kind shown in Figure 2.3.

For the hypothetical aircraft represented in Figures 2.2, 2.3 and 2.8 the maximum lift coefficient C_{Lmax} is 1.5 at very low Mach numbers and with a wing loading of 5000 N/m² the EAS at this lift coefficient is 74 m/s. At a height of 2 km the corresponding Mach number is only

0.25 and this is not likely to be high enough to have any significant effect on C_{Lmax}, but at a height of 10 km the Mach number for the same V_e of 74 m/s has risen to 0.425 and this is likely to cause some reduction of C_{Lmax}, perhaps to 1.25. For this value of C_L the corresponding EAS is recalculated as 81 m/s and the β curve in Figure 2.8 would then be approximately as shown by the broken line on the left-hand side of the figure. It should be emphasised that the numerical values given here are hypothetical and are intended only to illustrate the kind of change that may occur in the β curve at high altitude and low EAS.

The conditions required for flight with maximum fuel economy will be discussed in Chapter 7, but even without considering such detailed arguments it is clear that for good economy a subsonic civil aircraft should fly at a Mach number no greater than that corresponding to the start of the steep drag rise as shown in Figure 2.7. Referring to Figure 2.8, this means that in the case considered there V_e should be kept below about 240 m/s at $h = 2$ km and below about 140 m/s at $h = 10$ km. Figure 2.6 shows that both these values of V_e correspond to a Mach number of about 0.8. It is therefore reasonable to assume, for a preliminary approximate analysis of subsonic civil aircraft performance, that for any given height the only part of the curve relating β to V_e that needs to be considered is that part which is not affected by varying Mach number. The example of Figure 2.8 shows that at a height of 10 km the range of V_e to be considered is between about 85 and 140 m/s which appears initially to be a rather small range of speed, but in fact it corresponds to a range of true air speeds from 146 to 241 m/s and Mach numbers from 0.48 to 0.80. For military aircraft and for supersonic civil aircraft for which the required range of Mach number is much wider, the effects on β are far more significant and these will be considered in Chapter 10. In the intervening chapters it will be assumed, unless otherwise stated, that the relation between β and V_e is independent of height. This simplification is the main reason for developing the analysis in terms of the equivalent air speed V_e rather than the true air speed V.

2.5 Introduction to climbing performance

The discussion of climbing performance to be given here is only a preliminary introduction; the subject will be considered more fully in Chapter 4. It is assumed that for a typical aircraft the curve shown as a full line in Figure 2.9 represents the variation of β with V_e. The curve is similar to that shown in Figure 2.8 and within the range of validity of the assumptions already discussed it may be considered to be valid at all heights. The curve rises vertically at $V_e = V_{es}$ where the lift coefficient reaches the maximum value C_{Lmax}, but the effective

Introduction to climbing performance

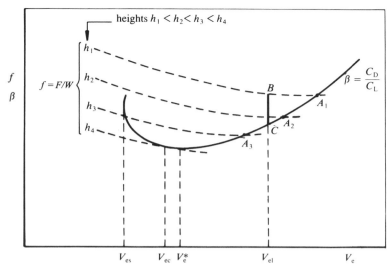

Figure 2.9. f and β as functions of V_e.

stalling speed as indicated to the pilot by the onset of buffeting or other phenomena would be somewhat greater than this speed and the minimum safe flying speed would be about 20% above the stalling speed, to allow an adequate margin of safety.

The broken lines in Figure 2.9 show the thrust/weight ratio $f = F/W$ for four heights at a specified weight. These are based on the maximum available thrust which decreases as the height increases from h_1 to h_4. For a given height, the variation of maximum available thrust with speed depends on the type of engine and will be discussed in Chapter 5. In many cases there is some decrease of thrust with increasing speed as shown in the figure, but for turbojet and turbofan engines the effect of varying speed may be small in the range of speed used for cruising and climbing and is often neglected in approximate performance calculations, so that the f curves in a figure such as this become a set of horizontal straight lines.

Equation (2.3) shows that if $\cos \gamma = 1$ the vertical distance between the f and β curves is equal to $\sin \gamma$, where γ is the angle of climb at maximum thrust. For example in Figure 2.9, at an EAS of V_{e1} and height h_1 the vertical intercept BC is equal to $\sin \gamma$. At the points A_1, A_2 and A_3 the maximum angle of climb is zero even at full thrust for the heights h_1, h_2 and h_3 and these points therefore give the maximum EAS in level flight.

The speed V_e^* shown in Figure 2.9 is the EAS for which β has the minimum value β_m. This corresponds to maximum aerodynamic efficiency because in steady level flight at this speed the necessary lift is generated with minimum drag and therefore with minimum thrust requirement. The speed V_{ec} shown just below V_e^* in Figure 2.9 is the EAS for maximum angle of climb, i.e. maximum $(f - \beta)$ and this requires some further explanation.

The set of broken curves in Figure 2.9 shows the progressive reduction in maximum thrust with increasing height, but a similar set of curves could be used to represent several fixed thrust settings below the maximum for a single height, say h_1. The progression from point A_1 downward to A_3 and lower would then show the steady level flight speeds at h_1 for successively lower thrust settings. At the lowest thrust setting for which level flight is possible at this height the f curve just touches the β curve, as the curve for height h_4 does in Figure 2.9, and the point where the curves touch gives the speed V_{ec}. The angle of climb has the maximum value of zero at this speed, the angle becoming negative if the aircraft is perturbed to a higher or lower speed without any change of thrust setting.

Taking the f curve which just touches the β curve as a datum, it may sometimes be possible to regard the other f curves, representing higher thrust settings or lower heights, as being 'parallel' to this datum curve. This means that each of the other curves can be derived from the datum curve by adding a constant increment of f which is independent of speed. The speed V_{ec} for maximum angle of climb is then independent of height or thrust setting, i.e. it is the same as the speed where the datum curve touches the β curve. In general this 'parallel' condition may not be exactly satisfied and there will then be some variation of V_{ec} with height and thrust setting.

At any height or thrust setting for which f is independent of speed the two speeds V_{ec} and V_e^* will be equal, but if f decreases with increasing speed as shown in Figure 2.9 the speed V_{ec} will be below V_e^*.

The *rate of climb* V_c is defined as the upward vertical component of velocity and is given by

$$V_c = dh/dt = V \sin \gamma = (V_e/\sigma^{1/2}) \sin \gamma \qquad (2.9)$$

and thus for a given height (fixed σ) the rate of climb is proportional to the product of V_e and the intercept BC appropriate to it, as seen in Figure 2.9. The conditions for maximum rate of climb will be examined in Chapter 4 and at this stage only a brief discussion will be given, based on consideration of energy.

In a climb at constant speed there is no change of kinetic energy and the energy added to the aircraft all appears as potential energy. The rate of increase of potential energy is WV_c and this must be equal to the difference between the thrust power FV and the rate at which

Introduction to climbing performance

energy is dissipated by drag DV. Thus

$$WV_c = V(F - D)$$

or

$$V_c = V(f - \beta), \quad \text{if} \quad \cos \gamma = 1, \quad \text{so that} \quad L = W. \tag{2.10}$$

This result, taken with Equation (2.3), confirms Equation (2.9) and the fact that the rate of climb is maximum when the product $V(f - \beta)$ is maximum. It is also clear that the maximum rate of climb varies with height, as the engine characteristic f has already been seen to decrease with increasing height.

In Equation (2.10) the speed V can be replaced by $V_e/\sigma^{1/2}$ and Figure 2.10 has been derived from Figure 2.9 to show the variation of βV_e with V_e. It remains to display the variation of fV_e similarly and although it would be easy to derive this from the broken curves of Figure 2.9 it is simpler for the sake of illustration to consider a turbojet or turbofan engine for which the maximum thrust at a given height can be assumed to be independent of speed. For this case the broken lines in Figure 2.9 would be horizontal and the product fV_e appears in Figure 2.10 as a family of straight lines from the origin, two of which are shown in the figure as OA and OB for two heights, OB being for the greater height. (Note that the horizontal tangent to the β curve in Figure 2.9 touches the curve where $V_e = V_e^*$ and $\beta = \beta_m$, and

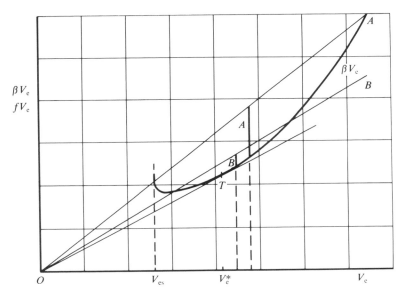

Figure 2.10. Derivation of maximum rate of climb.

this horizontal line converts to a tangent from the origin to the βV_e curve in Figure 2.10, touching at the point T where $V_e = V_e^*$.)

The vertical intercepts A and B have been drawn at the speeds where $V_e(f - \beta)$ is a maximum and since V_e is proportional to V, for any given height, these intercepts are proportional to the maximum rates of climb at the two heights. For this case where f is independent of speed Figure 2.9 would show that the maximum *angle* of climb occurred at $V_e = V_e^*$, whereas Figure 2.10 shows that the maximum *rate* of climb occurs at an EAS greater than V_e^*, with an angle of climb that is less than the maximum. The speed for maximum rate of climb increases as the available thrust increases, either because of reduced height (OA instead of OB) or because the throttle setting is higher as explained earlier. The extreme case of zero rate of climb (and zero angle) is represented by the point T in Figure 2.10, with $V_e = V_e^*$. In cases where f decreases with increasing speed, as shown by the broken curves in Figure 2.9, the speeds for maximum rate of climb would be less than those shown in Figure 2.10.

For propeller aircraft it is sometimes assumed that the shaft power transmitted to the propeller is independent of forward speed. Then if it can also be assumed that the propeller efficiency does not vary much with forward speed over the relevant range, the thrust power FV is roughly constant, for any given height, and Equation (2.10) shows that the maximum rate of climb is obtained at a speed giving minimum DV, i.e. minimum βV_e. In many cases this speed is close to the stalling speed, as in the example shown in Figure 2.10, so that in practice the useful speed for maximum rate of climb may be rather greater than the speed for minimum βV_e. Moreover, as will be explained in Chapter 5, increasing forward speed often gives an increase of shaft power and the propeller efficiency may also increase, especially in the lower part of the speed range. Both these changes lead to an increase of thrust power FV with forward speed, thus tending to increase still further the EAS for maximum rate of climb, although this speed will nearly always be less than V_e^*, in contrast to the cases illustrated in Figure 2.10 where the best speed is above V_e^*. The evaluation of the speed for minimum βV_e will be considered in Chapter 3.

When the rate of climb has been found for a range of heights, speeds and perhaps thrust settings the time required to climb between heights h_1 and h_2 can be found from the first part of Equation (2.9) as

$$t = \int_{h_1}^{h_2} \frac{1}{V_c} \, dh. \tag{2.11}$$

Essentially this requires an evaluation of V_c at a number of heights within the required range, but several options are open, depending on the flight conditions. Diagrams like Figures 2.9 and 2.10 can be used to find V_e for a particular angle of climb or rate of climb at any height and thrust setting, and if the total time to climb from h_1 to h_2 is required to be a minimum the speed should be adjusted continually during the climb to keep V_c at the maximum value for each height. The subject will be discussed further in Chapter 4.

2.6 Upper limits of height

The curve shown in Figure 2.9 for f at height h_4 just touches the β curve at $V_e = V_{ec}$, so that the angle of climb has a maximum value of zero at this speed and is negative at any other speed or at any greater height. The height h_4 is the greatest height attainable in steady flight and is known as the *absolute ceiling*. Flight at this height is possible only at the one speed where $V_e = V_{ec}$. As the weight of the aircraft decreases due to consumption of fuel the ratio $f = F/W$ increases, for any given height and speed, so that the height for which the f curve just touches the β curve also increases. Thus the absolute ceiling increases gradually as fuel is used. In practice the absolute ceiling is never quite reached and in flight very close to it any movement of the controls away from the exact trim position leads to a loss of height, either because the speed changes or because there is an increase of drag.

The low air density at very high altitudes means that a lift equal to the weight of the aircraft can be generated only by flying at a high true air speed and hence at a high Mach number. Taking the conditions of Figure 2.8 as an example, the EAS for minimum β at low Mach number is about 110 m/s, but Figure 2.6 shows that at this EAS the Mach number will be greater than 0.8 at any height above 13 km, so that C_D and β will be substantially increased. Even at the lower EAS of 100 m/s the Mach number at $h = 13$ km is about 0.73 and the maximum usable lift coefficient might then be less than the value of 0.82 which would be required for this wing loading of 5000 N/m^2. Thus the ceiling of an aircraft may be limited by the low air density at high altitudes and cannot be increased indefinitely by increasing the available thrust; the height that is attainable is limited by the adverse effects of high Mach numbers on C_D and C_{Lmax}. For flight at very high altitudes it is usually necessary to reduce the wing loading so that the necessary lift can be generated at lower values of V_e and M, although there are specialised aircraft such as the SR-71 shown in Figure 2.11 which can fly at very high altitudes at high supersonic speeds.

Since the rate of climb tends to zero as the absolute ceiling is approached a more useful indication of the practicable upper limit is the maximum height at which a specified rate of climb can be maintained in steady flight. In the early days of aviation the *service ceiling* was defined in terms of a rate of climb of 100 ft/min

26   *Basic flight theory*

($= 0.51$ m/s) and later the *operational ceiling* was defined for a rate of climb of 500 ft/min ($= 2.54$ m/s). For modern aircraft a rate of climb of 0.5 m/s is usually too small to form the basis of a useful definition, whereas 1.5 m/s defines a more useful upper limit of height and to avoid ambiguity it is suggested that this should be specified as 'the 1.5 m/s ceiling'. Using this form of definition it is simple to define alternative ceilings in terms of different maximum available rates of climb.

Another useful definition of a practical upper limit of height is the maximum height at which a specified form of turn can be maintained without loss of height or speed and this will be discussed in Chapters 8 and 10.

2.7   Further discussion of the speed V_{ec}

As already explained, V_{ec} is the EAS for maximum $(F - D)$, or maximum $(f - \beta)$, and gives the maximum angle of climb.

Figure 2.11. The Lockheed SR-71 Blackbird reconnaissance aircraft has an exceptionally high ceiling of more than 25 km, roughly double that of a typical jet transport.

Further discussion of the speed V_{ec}

Reference to Figure 2.9 and the f curve for height h_3 shows that within the range of speed giving positive angles of climb there are in general two alternative values of V_e for any one specified angle of climb, i.e. there is one speed above V_{ec} and another speed below it for which the intercepts between the f and β curves are the same. The conditions $V_e > V_{ec}$ and $V_e < V_{ec}$ define two distinct flight regimes, the first having speed *stability* and the second speed *instability*. It will now be shown how these two regimes differ in a return to steady equilibrium flight after a speed disturbance. A full analysis of speed stability or instability is beyond the scope of this book and reference may be made to Etkin (1972) and Neumark (1957) for further study, but the importance of the critical speed V_{ec} can be illustrated by considering one special case for which the analysis is quite simple.

Suppose that an ideal control system operates the elevator in such a way that the angle γ is kept constant, without any change of thrust setting, in the period following a speed disturbance until equilibrium flight is reestablished. Figure 2.12 shows an enlargement of a portion of Figure 2.9, the single engine characteristic supporting the observation that steady level flight at the height h_3 would normally be at speed V_{e3}. In contrast to this, flight at a fixed positive value of γ could be maintained at either of the two speeds V_e^+ or V_e^- for which equal intercepts are shown as upward arrows between the two curves.

Now suppose that from the equilibrium speed V_e^+ there is a positive speed perturbation of ΔV_e. To maintain the same intercept (rising to the point b) the thrust required would be greater than that available, thus there would be a thrust deficit and the speed would fall back to its original value. If the initial equilibrium flight had been at V_e^-,

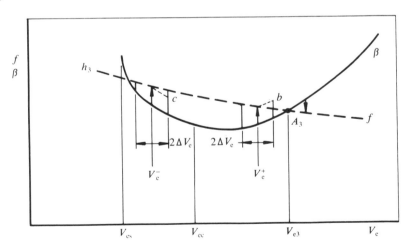

Figure 2.12. Intercepts used to explain speed stability.

maintenance of a constant intercept for the same positive speed perturbation ΔV_e (dropping to the point c) would require a thrust smaller than that available. There would then be an excess of thrust causing the speed to increase up to V_e^+, where equilibrium would again be established at the angle of climb γ.

If, in the first case, a speed decrement had been imposed from V_e^+ the consequent excess thrust would have restored the speed to V_e^+, whereas a speed decrement imposed from V_e^- would have led to a thrust deficit for this angle of climb, deceleration would have occurred and stalling would have been the inevitable consequence. It can be seen that in this special case, where a constant angle of climb is maintained by use of the elevator, departures in speed of either sign from V_e^+ are ultimately suppressed, thus displaying speed stability, whereas departures from V_e^- lead to further divergence of speed, showing speed instability.

It should be emphasised that the special case considered here has been introduced mainly because the analysis of the speed variation is then particularly simple. In practice a pilot does not usually maintain a constant climb angle solely by use of the elevator control; if a constant climb angle is required the engine throttle is adjusted as well as the elevator. It should therefore not be concluded that flight at an EAS less than V_{ec} is forbidden, although there can sometimes be difficulty, and even danger, particularly during a landing approach (with $\gamma < 0$). The arguments given above are still valid but in Figure 2.12 the f curve labelled for maximum available thrust at the height h_3 must be replaced by a similar one appropriate to a low altitude and low thrust setting. The intercepts for constant (negative) γ are downward, like the arrow shown at a speed above V_{e3}, but the particular intercept that is relevant is at a speed between the stalling speed V_{es} and V_{ec}. In these conditions of speed instability the pilot can maintain a glide path with a constant angle of descent at a constant speed only by continual adjustment of both the elevator and the thrust setting. This is difficult and possibly dangerous with all the controls operated manually, but the difficulty can be overcome by the use of an appropriate automatic control system, although Pinsker (1972) has drawn attention to the possibility of instability of the glide path if the engine thrust is controlled automatically while the elevator is operated manually. If both the thrust and the elevator are automatically controlled a straight glide path can be maintained at constant speed with no instability.

In discussions of speed stability the speed V_{ec} is often referred to as the 'minimum drag speed', but the true minimum drag speed (at constant lift) is V_e^* and this is not the same as V_{ec} except in the special case where the thrust at a given height is independent of speed and the

f curve becomes a horizontal line. The regime where $V_e < V_{ec}$ is sometimes said to be 'on the back of the drag curve'.

The form of speed stability that has been discussed here should not be confused with longitudinal static stability or pitch stiffness, which is sometimes called 'speed stability' because it is related to the change of speed that is ultimately established after a specified displacement of the elevator. With a statically stable aircraft flying initially along a level flight path at a speed V_{e3}, corresponding to the point A_3 in Figure 2.12, a downward deflection of the elevator will lead eventually to a new steady state at a higher speed. If there is no change of thrust setting the angle γ will then be negative and an increase of thrust will be needed to achieve level flight at the higher speed. If, from the same initial state A_3, the thrust is increased without any change of elevator angle the aircraft will climb without any change of speed (unless the thrust line is at some distance either above or below the CG so that a significant change of elevator angle is required to maintain zero pitching moment with the increased thrust). Thus the elevator is the primary control for speed and the thrust controls the angle of climb. For a further discussion of static stability, reference may be made to Babister (1980) and Etkin (1972).

3
Drag equations

The drag acting on an aircraft is of supreme importance in determining either the performance obtainable with a given thrust or the thrust required to achieve a specified performance, the latter being important because for a given speed the rate of consumption of fuel is approximately proportional to thrust. This chapter gives a brief account of the principal components of drag and the essential flow mechanisms on which they depend. Particular attention is paid to the nomenclature used for the drag components and to the conditions for various minima, because the undisciplined use of some of the terms has led to confusion in the past.

The drag polar for an aircraft has been discussed in Chapter 2 and in this chapter the representation of the polar by a simple mathematical expression is considered. The commonly used simple parabolic drag law has the great advantage of allowing many aspects of performance to be expressed in terms of simple equations, but the limitations of this drag law are emphasised and some alternative laws are introduced. In using either the simple parabolic law or any of the alternatives it is important to ensure that the constants in the equations are chosen to give the best possible agreement with the real drag polar over the range of C_L that is important.

The flight conditions for minimum drag and for minimum drag power have been discussed briefly in Chapter 2 and in this chapter they are examined in more detail, making use of the parabolic equation for the drag polar mentioned earlier.

3.1 Components of drag

The total drag of an aircraft may be regarded as the sum of several components. In one way of considering this the components are distinguished by the way in which force is exerted on the aircraft, either through the distribution of normal pressure or by the skin friction or shear stress at the surface. In the other approach the

Components of drag

components of drag are defined in terms of the fluid mechanisms that cause drag, with the aim of developing a rational scheme for the theoretical estimation of drag. This second approach is the one that is useful in developing equations for representing the drag polar and is the one to be considered here.

Figure 3.1 is based on a report of the Definitions Panel of the Aeronautical Research Council (1958) and shows the total drag divided into two parts, a lift-dependent drag represented by the coefficient C_{DL} and a drag which may be considered to be independent of lift. The latter component is then divided into two further parts, a datum drag and a spillage drag. A datum condition is defined in terms of a specified datum lift coefficient and a specified datum flow into the engines. Any change in engine intake flow (usually a reduction) from the datum may cause additional drag and this is the spillage drag. The additional drag resulting from any change of lift coefficient from the datum value is regarded as lift-dependent drag. The three components of datum drag shown in Figure 3.1, viscous drag, wave drag and vortex drag, all refer to the datum condition. Thus, for example, the total viscous drag is the sum of two components, one at the datum condition and another that is lift-dependent. Similarly the vortex drag and wave drag each have two components. The spillage drag is assumed to be independent of lift coefficient.

3.1.1 VORTEX DRAG

At subsonic speeds and in many cases also at supersonic speeds the most important component of lift-dependent drag is the part that is associated with the formation of a trailing vortex sheet behind the wing. This component of drag has already been discussed briefly in Chapter 2 and reference has been made to Clancy (1975). The name used for this component by Clancy and by numerous other authors is 'induced drag' but this name will not be used here because it is ambiguous; it has sometimes been used to mean the total lift-dependent drag. To avoid the ambiguity the name 'vortex drag' will be used as in Chapter 2.

The *aspect ratio* of a wing is of great importance in determining the vortex drag. This is defined as the ratio of the span b to the mean value of the chord, the distance between the leading and trailing edges. The mean chord is

$$\bar{c} = S/b,$$

where S is the area of the wing and the aspect ratio is

$$A = b/\bar{c} = b^2/S.$$

For wings with aspect ratio greater than about 3 the vortex drag can

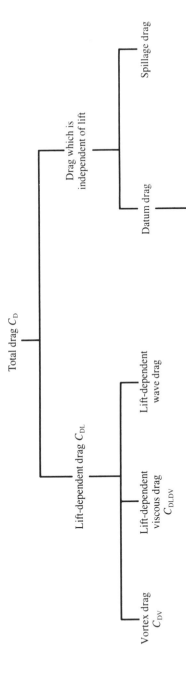

Figure 3.1. Components of drag.

be calculated with good accuracy by lifting-line theory, in which the lifting wing is represented by a lifting bound vortex extending along a spanwise line in the wing. The variation of the strength of this vortex along the span leads to the formation of a trailing vortex sheet behind the wing as mentioned in Chapter 2. In reality this vortex sheet rolls up to form two concentrated trailing vortices but in the simple theory this rolling up process is ignored. The theory is explained by Clancy (1975) and a more advanced account is given by Schlichting & Truckenbrodt (1979).

The lifting-line theory shows that for a given lift coefficient and aspect ratio the vortex drag coefficient C_{DV} is a minimum when the downwash velocity induced by the trailing vortex sheet is uniform over the span. This uniformity of downwash is obtained when the spanwise distribution of lift is elliptic and the vortex drag coefficient is then

$$C_{DV} = C_L^2/(\pi A). \tag{3.1}$$

More generally, when the spanwise lift distribution is not elliptic and the downwash velocity is not uniform, the vortex drag coefficient may be expressed as

$$C_{DV} = k_v C_L^2/(\pi A) \tag{3.2}$$

where k_v is the *vortex drag factor*, usually having a value between 1.0 and 1.1 which can be calculated from the lifting-line theory.

Küchemann (1978) has shown that these results are much more generally applicable than is implied by the limitations of lifting-line theory and that for *any* lifting wing with given C_L and A the vortex drag is a minimum when the downwash velocity is uniform at a station far downstream of the wing. (For the simple model assumed in lifting-line theory, uniformity of downwash far downstream also implies uniformity at the wing.) Küchemann shows that Equation (3.2) applies to any lifting wing and that for a non-planar lifting system, such as a wing with end-plates, it is possible for the factor k_v to be less than 1.

The datum lift coefficient, at which the datum drag is measured, may be specified in any way that is convenient but is usually chosen to be either zero or the value at which the total drag coefficient C_D is a minimum. Even if the datum C_L is chosen to be zero there may still be a small vortex drag at the datum condition, because on a twisted wing at zero overall lift there will be positive lift over some parts of the span and negative lift over others, so that trailing vortices will be formed and will cause some drag.

3.1.2 WAVE DRAG

Wave drag is associated with the formation of shock waves and occurs at high subsonic and supersonic speeds. At these speeds there is a component at the datum condition and also, in general, a lift-dependent wave drag. Except in Chapter 10 it will be assumed that the Mach number is low enough to keep the wave drag small and the lift-dependent component of it will be neglected.

3.1.3 VISCOUS DRAG

Viscous drag is associated with the development of the boundary layer over the external surface of the aircraft. Except when the wing is stalled, the greater part of the viscous drag is a direct consequence of the skin friction or shear stress at the surface, but there is also a component caused by the effect of the boundary layer on the distribution of normal pressure.

The lift and drag curves of Figure 2.2 are typical of an aircraft with a cambered wing section and show that the minimum in C_D occurs at a small positive value of C_L and is not related to any special feature of the lift curve; in particular it does not occur at zero lift. The same remark applies to the *viscous* drag coefficient which is usually minimum at a small positive lift coefficient and rises as the lift coefficient increases above this value.

3.1.4 SPILLAGE DRAG

The form of the streamlines approaching an engine intake depends on the flow rate through the engine and hence on the engine thrust. This is illustrated by the sketches of streamlines in Figure 3.2, where conditions of high and low thrust are shown. When the thrust is reduced from a high to a low value there is a decrease in the flow rate through the engine and more of the air approaching the intake 'spills' around the outside of the engine nacelle. There is then a change in the distribution of pressure over the external surface of the nacelle and in general this causes a change of drag. The spillage drag is defined as the difference between the drag at a given engine operating condition and that at a specified datum flow rate and in most cases it is positive when the flow rate is below the datum level.

3.2 Equations representing the drag polar

The drag polar as shown for a typical case in Figure 2.3 may be represented by an equation in several possible ways, but it is always essential that the equation should show the two principal components of drag, one dependent on lift and the other independent of lift. The datum conditions for lift and drag must also be implicit in the equation and in the simplest approach it is assumed initially, for the purpose of deriving the equation, that the datum lift coefficient is zero, while the datum drag coefficient is necessarily positive. The consequence is a total drag coefficient which is the sum of two terms, one constant and the other a function of C_L.

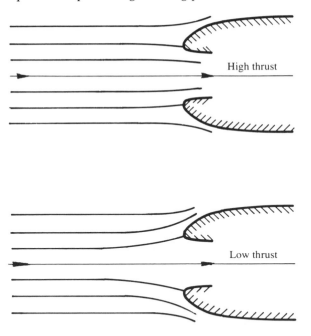

Figure 3.2. Engine intake flows.

With the lift-dependent wave drag neglected the total lift-dependent drag coefficient is

$$C_{DL} = C_{DV} + C_{DLDV} \tag{3.3}$$

where C_{DLDV} is the coefficient of lift-dependent viscous drag. The assumption is now made that the variation of C_{DLDV} with C_L can be represented by

$$C_{DLDV} = K_{LDV} C_L^2 \tag{3.4}$$

where K_{LDV} is a constant, normally positive. This has been shown by experiments to be a reasonably accurate representation for most aircraft except when the lift coefficient is either very small or so large that the wing is close to the stalling condition. As the stall is approached separation of the boundary layer causes a rapid rise of C_{DLDV} with increasing C_L. When C_L is very small Equation (3.4) becomes unsatisfactory because, as already mentioned, the total viscous drag coefficient is usually minimum at a small positive lift coefficient and not at zero lift. Thus with the datum lift coefficient chosen to be zero, C_{DLDV} and K_{LDV} have to be negative for small positive values of C_L.

For the present purpose of developing an equation to represent the drag polar it will be assumed that Equation (3.4) is valid with constant

K_{LDV}, even at small values of C_L. Then this equation, with Equations (3.2) and (3.3), shows that

$$C_{\text{DL}} = (k_v/(\pi A) + K_{\text{LDV}})C_L^2 \tag{3.5}$$

and if K_1 is the drag coefficient at the datum condition of zero lift the *total* drag coefficient is

$$C_D = K_1 + K_2 C_L^2 \tag{3.6}$$

where

$$K_2 = k_v/(\pi A) + K_{\text{LDV}}.$$

Following the nomenclature of ESDU 81026 (1986) Equation (3.6) will be called the *simple parabolic drag law*, although it is really only a useful empirical equation and not a statement of a physical law.

The presence of spillage drag when the engine flow departs from the datum condition is in conflict with the assumption made in Chapter 1 that the thrust and drag are independent and implies that K_1 is not strictly constant, but depends to some extent on the thrust of the engines. Nevertheless the assumption of constant K_1 gives a good approximation to the true drag polar in most cases because for normal flight conditions and thrust variations the spillage drag is usually small enough to be neglected in making early performance estimates. In cases where some allowance for spillage drag is considered to be necessary the drag increment can be replaced by an equivalent negative increment of thrust.

It has already been explained that Equation (3.4) becomes inaccurate when C_L is either very large or very small and this must also be true of Equation (3.6), which depends on (3.4). A further reason for inaccuracy of Equation (3.6) at very small C_L is that the vortex drag factor k_v is not strictly constant at low C_L if the wing is twisted. To overcome these difficulties the constants K_1 and K_2 should be chosen to give the best possible representation of the true drag polar over the range of C_L that is covered in normal cruising and climbing flight, and not at values of C_L close to zero. This is illustrated qualitatively in Figure 3.3, where the curve shown as a full line gives typical values of C_D plotted against C_L^2 and the broken straight line represents Equation (3.6), with K_1 and K_2 chosen to give values of C_D close to the correct ones over a wide range of C_L. It should be noted particularly that the value shown for K_1 is not the true value of C_D at $C_L = 0$, although this is suggested by the equation. The discrepancy at low values of C_L is exaggerated in the Figure and is not shown to scale.

It has been traditional in the literature of aerodynamics to express the constant K_1 in Equation (3.6) as C_{D0}, purposely intending it to

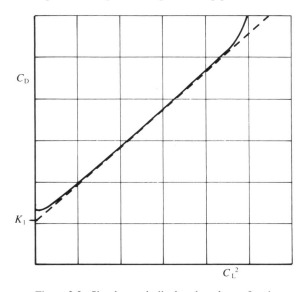

Figure 3.3. Simple parabolic drag law shown fitted to a real drag curve.

represent the drag coefficient at zero lift, as the equation implies. This notation is not used here because it can lead to confusion between the constant used in the equation and the true value of C_D at zero lift. There has also been a tradition of replacing K_2 in the same equation by $k/(\pi A)$ in order to produce a form similar to Equation (3.2). With this notation Equation (3.6) becomes

$$C_D = K_1 + kC_L^2/(\pi A). \tag{3.7}$$

The factor k has often been called the 'induced drag factor' or the 'lift-dependent drag factor' but neither of these names will be used here because both are ambiguous. 'Induced drag factor' has been used to mean either k_v or k and 'lift-dependent drag factor' has been used to mean either K_2 or k.

In Equation (3.3) the vortex drag coefficient C_{DV} is usually much greater than the viscous term C_{DLDV}, so that in Equation (3.5) the dominant term within the brackets is $k_v/(\pi A)$ and in Equation (3.6) the constant K_2 is roughly proportional to A^{-1}. Thus increase of aspect ratio has a large beneficial effect in reducing K_2, although in practice this has to be balanced against the increased weight of the wing structure that is needed for higher aspect ratios. An alternative to increasing aspect ratio is the use of winglets at the wing tips, sometimes known as wing tip fences. These were proposed by Whitcomb (1976) and are illustrated in Figure 3.4; they have an effect

38 Drag equations

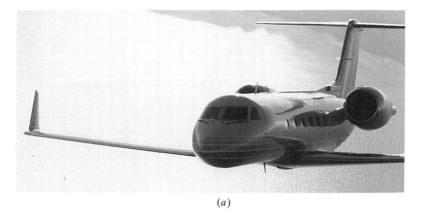

(a)

(b)

Figure 3.4. (a) The winglet used on the Grumman Gulfstream IV; (b) The wing tip fence used by Airbus Industrie.

Equations representing the drag polar

similar to that of increasing aspect ratio in reducing K_2, although they also give some increase of K_1 because of the extra surface area. A further account of the use of winglets has been given by Coppi (1980).

Typical values of K_1 for subsonic aircraft are in the range 0.012–0.025 for the cruising configuration, with undercarriage, flaps and slats all retracted. The value of $k(=\pi A K_2)$ usually lies in the range 1·15–1·4, but since

$$k = k_v + \pi A K_{LDV}$$

there is some tendency for k to increase as the aspect ratio becomes large.

An alternative notation which is sometimes used, especially in the USA, defines

$$\text{Efficiency factor} = e = 1/k,$$

so that Equation (3.7) becomes

$$C_D = K_1 + C_L^2/(\pi A e). \tag{3.8}$$

With this notation Ae may be regarded as an effective aspect ratio, defined so that the total lift-dependent drag coefficient C_{DL} is equal to the vortex drag coefficient for a wing of aspect ratio Ae having elliptic loading, giving $k_v = 1$ in Equation (3.2).

3.2.1 ALTERNATIVE PARABOLIC EXPRESSIONS

An alternative representation of the drag polar which is more accurate than Equation (3.6) at low values of C_L is the *modified parabolic drag law* described in ESDU 81026 (1986). This is based on the choice of the datum lift coefficient as the value C_{L_1} at which the total drag coefficient C_D is a minimum. The minimum drag coefficient is then denoted by K_3 and the drag coefficient at any value of C_L is expressed as

$$C_D = K_3 + K_4(C_L - C_{L_1})^2. \tag{3.9}$$

In the example of Figure 2.3 the minimum C_D occurs when C_L is about 0.23 and for low or moderate lift coefficients the rise of C_D as C_L either increases or decreases from this datum condition is better represented by Equation (3.9) than by (3.6). An alternative and exactly equivalent equation which leads to simpler analytical results is obtained by writing

$$K_5 = K_3 + K_4 C_{L_1}^2,$$

so that

$$C_D = K_5 + K_4 C_L^2 - 2K_4 C_L C_{L_1}. \tag{3.10}$$

The use of Equation (3.9) or (3.10) eliminates most of the discrepancy shown in Figure 3.3 for low values of C_L, but even so Equation (3.6) will be used in the following analysis because it is simpler and usually gives an adequate representation of the true drag polar over the range of C_L that is covered in cruising and climbing flight. It is important in all cases to choose the most suitable values of K_1 and K_2 for the range of C_L that is to be considered and it may sometimes be desirable to choose different values of K_1 and K_2 for different aspects of performance analysis for the same aircraft.

A formal statement of this concept of varying K_1 and K_2 to suit the range of C_L to be covered is embodied in the *dual-parabolic drag law* which is described in ESDU 81026 (1986). This is based either on Equation (3.6) or on (3.9) but two sets of values are used for the pair of coefficients, K_1 and K_2 or K_3 and K_4, one for the lower range of C_L and the other for the upper range.

3.3 Equations based on the simple parabolic drag law

Equation (3.6) representing the simple parabolic drag law shows that

$$\beta = C_D/C_L = K_1/C_L + K_2 C_L. \tag{3.11}$$

Any function of the form $Ax^n + Bx^{-n}$, where A, B and n are constants, has a minimum value when $Ax^n = Bx^{-n}$. Hence the minimum value of β occurs when

$$C_L = C_L^* = (K_1/K_2)^{1/2} \tag{3.12}$$

and the minimum value itself is

$$\beta_m = 2(K_1 K_2)^{1/2}. \tag{3.13}$$

For steady straight flight with $\cos \gamma = 1$ a useful alternative expression for β can be obtained by using Equation (2.4). This is

$$\beta = [\tfrac{1}{2}\rho_0 K_1/w] V_e^2 + [2w K_2/\rho_0] V_e^{-2} \tag{3.14}$$

and it is convenient to write it as

$$\beta = P V_e^2 + Q V_e^{-2}, \tag{3.15}$$

where P and Q vary with the wing loading w, for a given aircraft with fixed values of K_1 and K_2. The EAS V_e^* for minimum β is given by

$$P V_e^{*2} = Q V_e^{*-2} \tag{3.16}$$

so that

$$V_e^* = (Q/P)^{1/4} = (2w/\rho_0)^{1/2} (K_2/K_1)^{1/4}. \tag{3.17}$$

Equations based on the simple parabolic drag law

The EAS V_e^* and the corresponding true air speed $V_e^*/\sigma^{1/2}$ are often called 'minimum drag speeds' but it is important to note that these terms are correct only for straight flight with $\cos \gamma = 1$, i.e. when $L = W$. The speed V_e^* is defined as the EAS for minimum β when this condition is satisfied, but when γ is not small and $L = W \cos \gamma$, the EAS for minimum β is different because it depends on the relation between f, γ and β as given by Equation (2.3).

It is useful to introduce a speed ratio

$$v = V_e/V_e^*$$

so that Equation (3.15) becomes

$$\beta = Pv^2 V_e^{*2} + Qv^{-2} V_e^{*-2}.$$

Then since $PV_e^{*2} = QV_e^{*-2} = (PQ)^{1/2} = (K_1 K_2)^{1/2} = \frac{1}{2}\beta_m$, the general expression for the drag/lift ratio (with $\cos \gamma = 1$) becomes

$$\beta = \tfrac{1}{2}\beta_m (v^2 + v^{-2}). \tag{3.18}$$

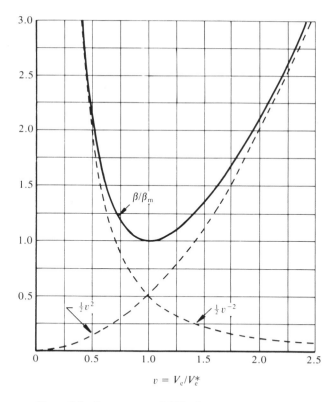

Figure 3.5. Components of β/β_m for straight and level flight.

Drag equations

An alternative notation which is sometimes used defines

$$n_1 = 1/v^2 = C_L/C_L^*.$$

(The symbol n is more commonly used but n_1 is used here to avoid confusion with the ratio L/W used in Chapters 8 to 10.) With this notation

$$\beta = \tfrac{1}{2}\beta_m(n_1 + 1/n_1) \tag{3.19}$$

and of course all the results to be derived later in terms of v can be expressed in terms of n_1 if required.

The three Equations (3.11), (3.15) and (3.18) all show β as the sum of two terms. The first term is

$$K_1/C_L = PV_e^2 = \tfrac{1}{2}\beta_m v^2 \tag{3.20}$$

and the second is

$$K_2 C_L = QV_e^{-2} = \tfrac{1}{2}\beta_m v^{-2}. \tag{3.21}$$

To illustrate this, Figure 3.5 shows the two components of β/β_m as given by Equation (3.18), together with their sum. It should be remembered of course that the curves become unrealistic when v is either so low that stalling occurs or so large (especially at high altitudes) that the high Mach number causes an increase of C_D.

EXAMPLE 3.1. RATE OF CLIMB

The drag polar of an aircraft can be represented by the simple parabolic drag law with $K_1 = 0.018$ and $K_2 = 0.045$. Find the values of β_m, C_L^* and V_e^* when the wing loading is 5 kN/m². Also find the rate of climb at a true air speed of 200 m/s and a height of 5 km if the ratio of thrust to weight is 0.2.

From Equations (3.12) and (3.13),

$$C_L^* = (K_1/K_2)^{1/2} = 0.6325$$

and

$$\beta_m = 2(K_1 K_2)^{1/2} = 0.0569.$$

With $\cos \gamma = 1$, the wing loading becomes

$$w = \tfrac{1}{2}\rho_0 V_e^{*2} C_L^* = 5000 \text{ N/m}^2$$

and Appendix 2 gives

$$\rho_0 = 1.225 \text{ kg/m}^3,$$

so that

$$V_e^* = [2w/(\rho_0 C_L^*)]^{1/2} = 113.6 \text{ m/s}.$$

Power required to overcome drag

At the specified speed and height

$$V_e = V\sigma^{1/2} = 200(0{\cdot}6009)^{1/2}$$
$$= 155.0 \text{ m/s}.$$

Then $v = 155.0/113.6 = 1.365$ and from Equation (3.18)

$$\beta = \tfrac{1}{2}\beta_m(v^2 + v^{-2}) = 0.0683.$$

Alternatively, β may be found by putting $V_e = 155.0$ m/s in the equation

$$\tfrac{1}{2}\rho_0 V_e^2 C_L = w$$

so that, with the use of other values given above, $C_L = 0{\cdot}3398$ and $\beta = K_1/C_L + K_2 C_L = 0.0683$.

The angle of climb is given by

$$\sin \gamma = f - \beta = 0.2 - 0.0683 = 0.1317.$$

This shows that $\cos \gamma = 0.991$ and justifies the assumption that $\cos \gamma = 1$. The rate of climb is

$$V_c = V \sin \gamma = 26.3 \text{ m/s}.$$

3.4 Power required to overcome drag

The power required to overcome drag is the *drag power DV* and the speed at which this is a minimum has already been discussed briefly in Chapter 2.

If $\cos \gamma = 1$, $DV = \beta W V_e / \sigma^{1/2}$ \hfill (3.22)

and for given W and σ this is minimum when βV_e is minimum. Equation (2.4) shows that

$$V_e \propto C_L^{-1/2}$$

so that

$$\beta V_e = V_e C_D / C_L \propto C_D / C_L^{3/2}$$

and this must be minimum for minimum drag power. It should be noted that this result is independent of any assumptions about the drag polar or its representation by an equation. Minimum drag power for an aircraft at any specified height is always obtained by adjusting the speed to make $C_D/C_L^{3/2}$ a minimum.

Using the simple parabolic drag law to represent the drag polar, Equation (3.18) shows that βV_e is minimum when

$$(v^3 + v^{-1}) \text{ is minimum,}$$

i.e. when

$$v = 3^{-1/4} = 0.76.$$

44   Drag equations

As mentioned earlier, this speed may often be close to the stalling speed and in this case the assumption that K_1 and K_2 are independent of C_L will be incorrect and the true value of v for minimum DV will be greater than 0.76.

As explained in Chapter 2, the condition for minimum DV is of interest in giving a rough indication of the speed for maximum rate of climb for a propeller aircraft. The condition also gives the speed for minimum rate of descent (rate of sink) in a glide with zero thrust, because the rate of loss of potential energy must then be equal to the drag power. The speed for minimum rate of sink should not be confused with the speed for minimum gliding angle which is given by $v = 1$, i.e. $\beta = \beta_m$.

Equations (3.18) and (3.22) show that in general the drag power is equal to

$$\tfrac{1}{2}\beta_m(v^2 + v^{-2})WvV_e^*\sigma^{-1/2}.$$

Thus the ratio of the drag power at any value of the speed ratio v to the minimum drag power is

$$[v^3 + v^{-1}][(0.76)^3 + (0.76)^{-1}]^{-1} = 0.570(v^3 + v^{-1}).$$

In level flight the minimum value of the required thrust power FV is equal to the minimum drag power and with v fixed at the optimum value Equation (3.22) shows that this is proportional to

$$\beta_m WV_e^*/\sigma^{1/2}.$$

Thus the minimum thrust power increases with height.

EXAMPLE 3.2. MINIMUM DRAG POWER

For the aircraft considered in Example 3.1, with a total mass of 100 000 kg, find the minimum drag power (a) at sea level and (b) at a height of 12 km. Also find for each height the drag force and the true air speed for the condition of minimum drag power.

For minimum drag power

$$v = 3^{-1/4} = 0.76$$

and

$$\beta = \tfrac{1}{2}\beta_m(3^{-1/2} + 3^{1/2}) = 1.155\beta_m = 0.0657.$$

From Equation (3.22) the minimum drag power is DV with

$$D = \beta W \quad \text{and with} \quad V = V_e/\sigma^{1/2} = vV_e^*/\sigma^{1/2}.$$

Hence

$$D = 0.0657 \times 10^5 \times 9.81 \times 10^{-3} \text{ kN} = 64.45 \text{ kN}$$

and this is independent of height.

The true air speed is $V = 0.76 \times 113.6\sigma^{-1/2} = 86.34\sigma^{-1/2}$ m/s. At sea level ($\sigma = 1$), $D = 64.45$ kN, $V = 86.34$ m/s and thus $DV = 5565$ kW, whereas at $h = 12$ km ($\sigma = 0.2537$) the drag remains the same at 64.45 kN while V and DV are almost doubled to $V = 171.42$ m/s and $DV = 11\,048$ kW, because at $h = 12$ km the density ratio gives $\sigma^{1/2} \simeq \frac{1}{2}$.

4

Climbing performance

In the design of a civil aircraft the condition of steady level cruise is of prime importance because improved fuel economy in this flight regime makes a direct and valuable contribution to the reduction of operating costs. Performance in the climb is often less important, but it cannot be ignored because a climb is always needed to reach the required cruising height after take-off and Air Traffic Control may also require the aircraft to change height during the cruise. For military aircraft, performance in the climb may be a primary design requirement because there is often a need to reach a specified height and speed in the shortest possible time, either from take-off or from some other prescribed initial conditions of height and speed.

The quantities that are of most interest in calculations of climbing performance are the rate of climb $V_c = V \sin \gamma$ and the time required and fuel used in climbing from one specified height to another. In many cases there is a change of speed during the climb, so that the aircraft is accelerating, but it will be shown that a correction can easily be made for the effect of the acceleration on the rate of climb. The angle of climb is also of some interest, although it is important mainly at low altitudes where there may be obstacles to be cleared or where a large angle of climb may be required for reasons of noise abatement.

It has been shown in Chapter 2 that the speed required to achieve either maximum rate of climb or maximum angle of climb depends not only on the drag polar but also on the way in which the maximum available thrust varies with flight speed. In this chapter the speeds required for maximum rate of climb and for maximum angle of climb will be considered in more detail, for various relations between maximum available thrust and flight speed. The simple conditions of constant F and constant FV will be considered first, followed by the more generally applicable relation $FV^n = $ constant. It will be assumed initially that the aircraft climbs at a constant true air speed, i.e. with zero acceleration, but effects of acceleration will be considered in § 4.5 and 4.6.

Before proceeding with detailed analysis of climb performance it is necessary to consider the approximation made in earlier chapters that the angle of climb γ is small enough to justify the assumption that $\cos \gamma = 1$. It will be shown in the following section that even when γ is not small the assumption that $\cos \gamma = 1$ introduces only small errors in the calculation of γ and V_c, for a climb with zero acceleration and with given values of the speed and height.

4.1 The approximation $\cos \gamma = 1$

When $\cos \gamma$ is not assumed to be equal to 1 the equation for β in terms of V_e is obtained from Equations (2.4) and (3.11) as

$$\beta = [\rho_0 K_1/(2w \cos \gamma)]V_e^2 + [(2wK_2 \cos \gamma)/\rho_0]V_e^{-2}. \tag{4.1}$$

The special equivalent air speed V_e^* which is derived from Equation (3.14) or (3.15) is defined as the EAS for minimum β when $L = W$ and is therefore always given correctly by Equation (3.17), even when the equations are being applied to a situation where $\cos \gamma \ne 1$ and $L \ne W$. Equation (3.13) also remains valid when $\cos \gamma \ne 1$ and substitution of Equations (3.13) and (3.17) in Equation (4.1), with $V_e = vV_e^*$, gives

$$\beta = \tfrac{1}{2}\beta_m(v^2 \sec \gamma + v^{-2} \cos \gamma). \tag{4.2}$$

Thus Equations (4.1) and (4.2) replace the Equations (3.14) and (3.18) as the more general forms of equation for the drag/lift ratio β.

The equation giving the corrected angle of climb at constant speed is then found by substituting Equation (4.2) for β in Equation (2.3), giving

$$\sin \gamma = f - \tfrac{1}{2}\beta_m(v^2 + v^{-2} \cos^2 \gamma). \tag{4.3}$$

Since the rate of climb is $V_c = V \sin \gamma$ it is directly proportional to $\sin \gamma$ and if γ_1 is the approximate angle of climb calculated by assuming that $\cos \gamma = 1$, the equation for $\sin \gamma_1$ is obtained from Equation (4.3) by putting $\cos^2 \gamma = 1$ in the last term. Then the relative error imposed by the approximation can be obtained from

$$(\sin \gamma - \sin \gamma_1)/\sin \gamma = \tfrac{1}{2}\beta_m v^{-2} \sin \gamma \tag{4.4}$$

and the percentage error in the calculated rate of climb for a given EAS is thus $100(\beta_m/2)v^{-2} \sin \gamma$. Also, if the error $(\gamma - \gamma_1)$ in the angle of climb is small,

$$\begin{aligned}\gamma - \gamma_1 &= (\sin \gamma - \sin \gamma_1)/\cos \gamma \\ &= \tfrac{1}{2}\beta_m v^{-2} \sin \gamma \tan \gamma \quad \text{(radians)}\end{aligned} \tag{4.5}$$

and the error is available directly, without solving Equation (4.3).

Figure 4.1 shows the error in angle of climb and Figure 4.2 shows the error per cent in rate of climb, both of these being plotted against

48   *Climbing performance*

γ for several values of the speed ratio v. The values shown in Figure 4.1 easily justify the approximation made in deriving Equation (4.5). The curves are calculated for $\beta_m = 0.06$, a typical value for a subsonic civil aircraft, but since both the errors are directly proportional to β_m the errors for other values of β_m can easily be derived. The errors are remarkably small, especially for the higher values of the speed ratio v, but it should be noted that the effects of high Mach number on C_D will make the curves for $v = 3$ unrealistic unless both the altitude and the wing loading are low. It will be shown in § 4.2 that if the thrust can be assumed to be independent of speed and if it is sufficient to achieve a

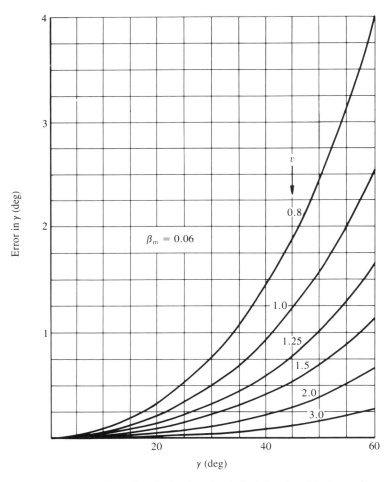

Figure 4.1. Error in calculated angle of climb for given V_e, due to the assumption that $\cos \gamma = 1$.

The approximation cos γ = 1

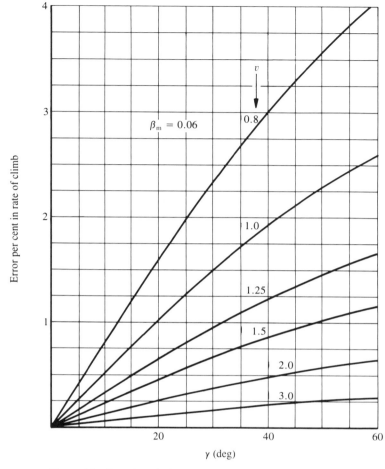

Figure 4.2. Error in calculated rate of climb for given V_e, due to the assumption that cos γ = 1.

large angle of climb, the value of v required for maximum rate of climb is substantially greater than 1. Figures 4.1 and 4.2 show that even for a climb angle of 50° the errors are then only of order 1° in angle and 1% in rate of climb.

EXAMPLE 4.1. EFFECTS OF ASSUMING THAT cos γ = 1

An aircraft has a wing loading of 4 kN/m² and a drag coefficient given by Equation (3.6) with $K_1 = 0.02$ and $K_2 = 0.05$. Assuming that cos γ = 1, calculate the angle of climb and the rate of climb at a height of 2 km and a true air speed of 170 m/s if the thrust/weight ratio f is 0.85. Repeat the calculations making the correct allowance for the value of cos γ.

In the approximate analysis β is given by Equation (3.18) and the angle of climb is found from Equation (2.3) with cos $\gamma = 1$, whereas the exact values of β and γ are obtained from Equation (4.2) and the complete Equation (2.3). For both forms of analysis some preliminary calculations are needed including, for $h = 2$ km at which $\sigma = 0.8216$,

$$V_e = V\sigma^{1/2} = 154.1 \text{ m/s}$$

and from Equation (3.17) the special equivalent air speed is

$$V_e^* = (2w/\rho_0)^{1/2}(K_2/K_1)^{1/4} = 101.6 \text{ m/s},$$

which leads to

$$v = V_e/V_e^* = 1.5164.$$

Equation (3.13) gives the related minimum value

$$\beta_m = 2(K_1 K_2)^{1/2} = 0.06325.$$

Then the approximate analysis gives

$$\beta = \tfrac{1}{2}\beta_m(v^2 + v^{-2}) = 0.08647$$

and

$$\sin \gamma_1 = f - \beta = 0.7635,$$

from which is obtained $\gamma_1 = 49.78°$ and with the true air speed of 170 m/s the rate of climb is found to be $V_c = 129.8$ m/s. The exact analysis gives

$$\beta = 0.07272 \sec \gamma + 0.01375 \cos \gamma$$

and an insertion of the approximate value $\gamma = 49.78°$ into Equation (2.3) gives

$$\sin \gamma = f - \beta \cos \gamma = 0.7715,$$

from which is obtained $\gamma = 50.49°$ and $V_c = 131.16$ m/s. Only one further iteration, using the new value of γ in Equations (2.3) and (4.2), yields

$$\sin \gamma = f - \beta \cos \gamma = 0.7717.$$

Then $\gamma = 50.51°$ and $V_c = 131.19$ m/s, these last two results being accurate to four figures.

The errors $(\sin \gamma - \sin \gamma_1)/\sin \gamma$ and $(\gamma - \gamma_1)$ are in agreement with the values calculated from Equations (4.4) and (4.5). The errors are a little greater than those shown in Figures 4.1 and 4.2 because in this example β_m is slightly greater than the value 0.06 used in deriving the figures. It is remarkable that the errors in γ and V_c are quite small, even though the correct value of β is 0.1231, 42% greater than the value found for $\cos \gamma = 1$ and the correct lift coefficient is $(2w \cos \gamma)/(\rho_0 V_e^2)$, 36% less than the value found for $\cos \gamma = 1$.

4.2 Climb of aircraft with thrust independent of speed

For approximate analysis of climbing performance it is often assumed that the maximum available thrust F at a given height is independent of flight speed, within the range of speed considered for high rates of climb. This assumption will be made here as the basis for a simple analysis, but it will be shown in § 5.2.2 and 5.2.4 that increasing speed usually gives a modest reduction of F for a civil turbofan at the climb rating and an increase of F for a typical military turbofan.

Equation (2.10) shows that if $\cos \gamma = 1$ the rate of climb at constant speed is

$$V_c = (V_e/\sigma^{1/2})(f - \beta) \tag{4.6}$$

and then if the thrust/weight ratio f is combined with the minimum value of β to define

$$\epsilon = f/\beta_m,$$

the substitutions $V_e = vV_e^*$ and $\beta = \tfrac{1}{2}\beta_m(v^2 + v^{-2})$ give

$$V_c = [V_e^* \beta_m/\sigma^{1/2}][v\epsilon - \tfrac{1}{2}(v^3 + v^{-1})]. \tag{4.7}$$

Assuming that F, f and ϵ are constant, the rate of climb has the maximum value $V_{c\,max}$ when v satisfies the equation

$$2\epsilon - 3v^2 + v^{-2} = 0, \tag{4.8}$$

which gives

$$v = \{\tfrac{1}{3}[\epsilon + (\epsilon^2 + 3)^{1/2}]\}^{1/2} \tag{4.9}$$

and when ϵ is large

$$v \simeq (2\epsilon/3)^{1/2}. \tag{4.10}$$

An alternative to Equation (4.8), using C_L as the variable instead of v, can be obtained from Equations (2.4), (3.11) and (4.6). This equation is

$$K_2 C_L^2 + fC_L - 3K_1 = 0 \tag{4.11}$$

and the positive root gives the lift coefficient for maximum rate of climb.

Using the variable v the rate of climb V_c is given by Equation (4.7), with V_e^* obtained from Equation (3.17). V_c has the maximum value when v is given by Equation (4.9) and Equations (2.3) and (3.18) show that the angle of climb is then given by

$$\sin \gamma = \beta_m [\epsilon - \tfrac{1}{2}(v^2 + v^{-2})], \qquad (4.12)$$

but it should be noted that this does not give the maximum angle of climb.

Figure 4.3 shows the speed ratio v and the angle of climb γ for maximum rate of climb. The ratio $\epsilon = f/\beta_m$ has been used for the abscissa, rather than the thrust/weight ratio f, because Equation (4.9) shows that v is a simple function of ϵ. The range of values of ϵ to be considered can be seen by noting that

$$\epsilon = (F/W)/(D/L)_{min},$$

so that in level flight with $F = D$ and $L = W$ the ratio ϵ is of order 1 if β ($=D/L$) is not far from the minimum value. Larger values of ϵ require large values of f but for transport aircraft f does not usually exceed 0.3, even at low altitudes, giving an upper limit of $\epsilon = 5$ if $\beta_m = 0.06$. Advanced training aircraft such as the BAe Hawk shown in

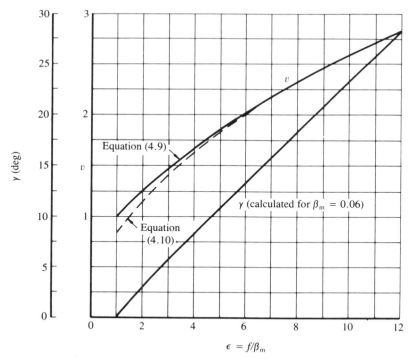

Figure 4.3. Conditions for maximum rate of climb, with thrust independent of speed.

Climb of aircraft with thrust independent of speed

Figure 4.4. The British Aerospace Hawk, a single-jet trainer with a thrust/weight ratio f of order 0.4 at low altitudes, can sustain a climb angle approaching 20° in steady flight. (In a decelerating climb the angle can be greater.) The angle that is evident here is the inclination of the aircraft axis to the horizontal, roughly 33°, and Figures 1.1 and 2.1 show that this is rather greater than the climb angle.

Figure 4.4 have somewhat larger values of f at low altitudes and for combat aircraft f may sometimes be as large as 1, although for these aircraft β_m is usually greater than that of a transport, typically about 0.085. With $f = 1$ and this value of β_m the ratio ϵ is nearly 12 and this may be taken as a rough guide to the upper limit of ϵ.

The full line curve for v in Figure 4.3 is calculated from the exact Equation (4.9) and the broken line from the approximate Equation (4.10), the difference between the two curves becoming insignificant when ϵ exceeds about 6. For the calculation of γ it has been assumed that $\beta_m = 0.06$, but values of γ for other values of β_m can easily be derived from the curve shown here, since Equation (4.12) shows that $\sin \gamma \propto \beta_m$. With $\cos \gamma = 1$ the lift is equal to the weight and ϵ is the ratio of the thrust to the minimum drag force. No climb is possible if this ratio is less than 1 and for the limiting case where $\epsilon = 1$ the maximum rate of climb is zero and is obtained at $v = 1$. The increase of v seen in Figure 4.3 as ϵ rises above 1 is in agreement with the qualitative discussion based on Figure 2.10 that has been given in § 2.5.

54   Climbing performance

Increasing height leads to a reduction in the maximum rate of climb, primarily because of the reduction in the maximum available thrust which will be discussed in Chapter 5. Equation (4.7) shows that for a given value of v the variation of V_c with height is also influenced directly by the change of σ, but the change of f (and hence of ϵ) is the dominant effect.

Since typical values of V_c^* are in the region of 100 m/s, Figure 2.6 shows that the high values of v shown in Figure 4.3 for large values of ϵ become unrealistic at high altitudes, because the Mach numbers are then so large that the drag equations used here (with constant K_1 and K_2) are no longer valid. When allowance is made for the rise of C_D and β at high Mach numbers the results shown for large ϵ are shifted to show a lower value for v and a higher value for γ. The speed for maximum rate of climb is then usually determined by the need to keep the Mach number below the value M_D at which the steep rise in C_D starts and the Mach number should be kept constant during the climb at an appropriate value which satisfies this condition. (For $h \leq 11$ km in the ISA there is a negative acceleration in a climb at constant Mach number, because of the change in the velocity of sound with height, and as will be explained in § 4.6 this leads to an increase in the rate of climb because the kinetic energy lost during the climb is available to increase potential energy.)

It has already been explained that for a military aircraft with $f \approx 1$ the upper limit of ϵ would be in the region of 12. Even larger values of f are possible, giving greater values of ϵ, but these have not been included in Figure 4.3 because the calculated values of v would be so large that the results would be unrealistic even at low altitudes. For an aircraft with $f > 1$ a steady vertical climb is possible, but for such an aircraft the energy equations given in § 4.5 should be used because the acceleration is usually large and the changes of kinetic and potential energy are often of the same order. Calculations for zero acceleration based on the drag equations used here show that the rate of climb is usually increased by reducing the angle of climb below 90°, but when correct allowance is made for the effects of high Mach number on drag it may often be found that the best rate of climb is achieved with a climb angle close to 90°.

The Equations (4.12) and (4.7) are based on the assumption that $\cos \gamma = 1$. If this assumption is not valid Equation (4.3) shows that these equations must be replaced by

$$\sin \gamma = \beta_m [\epsilon - \tfrac{1}{2}(v^2 + v^{-2} \cos^2 \gamma)] \tag{4.13}$$

and

$$V_c = [V_c^* \beta_m / \sigma^{1/2}][v\epsilon - \tfrac{1}{2}(v^3 + v^{-1} \cos^2 \gamma)] \tag{4.14}$$

Climb of aircraft with thrust power independent of speed

and for maximum rate of climb the speed ratio v should have the value required to make

$(2\epsilon v - v^3 - v^{-1}\cos^2\gamma)$ a maximum.

Calculations for the typical value $\beta_m = 0.06$ show that the values of v required to satisfy this condition do not differ by more than 0.2% from those found by using Equation (4.9), for all values of ϵ up to 20, even though γ for maximum rate of climb at $\epsilon = 20$ is as large as 53°. For the same value of β_m the error in the calculated value of γ for maximum rate of climb, due to the use of Equation (4.12) instead of (4.13), is also very small, being only about 0.15° when $\epsilon = 15$ and even less for smaller values of ϵ. Obviously the small errors in v and γ mean that the error in the calculated rate of climb is also small. The reason why there are only small errors in γ and in the optimum value of v is that when γ is not small both ϵ and v are large, so that $v^{-2}\cos^2\gamma \ll \epsilon$ and the $\cos^2\gamma$ terms in Equations (4.13) and (4.14) are negligible.

EXAMPLE 4.2. MAXIMUM RATE OF CLIMB

An aircraft has a wing area of 500 m² and a drag coefficient given by $K_1 = 0.018$ and $K_2 = 0.043$. If the maximum available thrust for climbing flight is independent of speed and is 300 kN at $h = 3$ km and 200 kN at $h = 9$ km, find the maximum rate of climb at constant speed at each of these heights when the total mass of the aircraft is 300 000 kg. Also find the angle of climb and the true air speed for the condition of maximum rate of climb.

Some preliminary calculations are needed, (i) to find ϵ, β and $\sin\gamma$, with $\beta_m = 2(K_1 K_2)^{1/2} = 0.05564$, and (ii) to find the special value of the equivalent air speed V_e^*. The wing loading is

$$w = W/S = 3 \times 10^5 \times 9.81/500 = 5886 \text{ N/m}^2.$$

Then Equation (3.17) gives $V_e^* = 121.9$ m/s.

Table 4.1 shows progress towards the three quantities required, for the two heights.

4.3 Climb of aircraft with thrust power independent of speed

For an aircraft with propellers driven by piston engines the shaft power is almost independent of forward speed, if the propeller speed is kept constant by automatic adjustment of the pitch and if the flight speed is low, as is usual for this class of aircraft in a climb. Then the thrust power FV can be assumed to be independent of flight speed if variations of propeller efficiency can be neglected. The same assumption of constant FV has sometimes been made for aircraft powered by turboprops, but it will be shown in § 5.4.1 that this is not usually correct; an increase of FV with speed can be expected for a modern turboprop. Nevertheless the assumption of constant FV is a

Table 4.1. *Calculations for Example 4.2*

h km	3	9
σ	0.7421	0.3807
f	0.1019	0.067 96
ϵ	1.8314	1.2214
v (from Equation (4.9))	1.205	1.055
$\beta = \frac{1}{2}\beta_m(v^2 + v^{-2})$	0.0595	0.0560
$\sin \gamma = f - \beta$	0.0424	0.0120

Then the true air speed is found from the definitions of v and V_e as

$V = vV_e^* \sigma^{-1/2}$ m/s	170.5	208.5

and because the values of v given above were calculated to give maximum V_c the last two sets of results can be combined to give

$V_{c\,max} = V \sin \gamma$ m/s	7.23	2.50

and finally from $\sin \gamma$ the two angles of climb are found as

γ	2.43°	0.69°.

Note the reduction of rate of climb with increasing height, due to reduction of thrust.

simple special case and its consequences will be examined in this section.

With constant FV, Equation (2.10) shows that maximum rate of climb is obtained at the speed giving minimum drag power DV. As shown in § 3.4 this is the speed for minimum $C_D/C_L^{3/2}$ (if $\cos \gamma = 1$) and if the simple parabolic drag law is valid the speed is given by $v = 0.76$. In many cases this speed will be close to or even below the stalling speed, so that a higher speed is required for safety and in any case the simple parabolic drag law with constant K_1 and K_2 may not be valid. Nevertheless it will be assumed here that the simple parabolic drag law is valid and that the condition $v = 0.76$ gives a sufficient margin of speed above the stalling speed. Then since FV is constant at any given height FV_e is also constant. If F_1 is defined as the thrust when $V_e = V_e^*$ (i.e. $v = 1$) and $f_1 = F_1/W$ the thrust F at any speed is given by

$$FV_e = FvV_e^*$$

Climb of aircraft with thrust power independent of speed

together with
$$FV_e = F_1 V_e^*,$$
so that
$$F = F_1/v \text{ and therefore } f = f_1/v.$$

The angle of climb is then given by
$$\sin \gamma = f - \beta = f_1/v - \tfrac{1}{2}\beta_m(v^2 + v^{-2}) \tag{4.15}$$
and with $v = 0.76$
$$\sin \gamma = f_1/0.76 - 1.155\beta_m. \tag{4.16}$$

The true air speed is $V = 0.76 V_e^*/\sigma^{1/2}$ and the maximum rate of climb is
$$\begin{aligned}V_{c\,max} = V \sin \gamma &= (V_e^*/\sigma^{1/2})(f_1 - 0.878\beta_m) \\ &= (V_e^* \beta_m/\sigma^{1/2})(\epsilon_1 - 0.878),\end{aligned} \tag{4.17}$$
where $\epsilon_1 = f_1/\beta_m$.

With increasing height the maximum thrust power FV decreases and since FV is equal to
$$V(F_1/v) = V_e^* F_1/\sigma^{1/2} = W V_e^* \epsilon_1 \beta_m/\sigma^{1/2},$$

Equation (4.17) shows that the maximum rate of climb also decreases. At the absolute ceiling $V_{c\,max} = 0$ and ϵ_1 must be equal to 0.878.

The Equations (4.16) and (4.17) and the conclusion that $v = 0.76$ for maximum rate of climb all depend on the assumption that $\cos \gamma = 1$. If this assumption is not valid it can be shown from Equations (2.1) and (2.3) that the condition required for maximum rate of climb is that $v\beta \cos \gamma$ should be minimum and Equation (4.2) shows that this requires
$$(v^3 + v^{-1} \cos^2 \gamma) \text{ to be minimum.}$$

Equation (4.13) can be used to show that values of $\cos^2 \gamma$ that are less than 1.0 reduce the value of v for maximum rate of climb below the special value of 0.76 and the amount of this reduction increases with thrust, i.e. with angle of climb. For example when $\beta_m = 0.06$ the value of v for maximum rate of climb is about 0.72 when the thrust is sufficient to give a climb angle of 15° at this speed, but the rate of climb is not sensitive to the exact value of v and if the climb is made at $v = 0.76$ instead of 0.72 the rate of climb is only reduced by about 1.5%. It was stated earlier that the low speed given by $v = 0.76$ may well be dangerously close to the stalling speed, or even below it, so these reduced values of v are even more likely to be unrealistic and it must be concluded that for the case where FV is constant the results of calculations based on the simple parabolic drag law are of little practical significance when the thrust is large enough to give an angle of climb greater than about 15° at $v = 0.76$. For these greater angles the speed used in practice to give maximum rate of climb will be the lowest that is compatible with safety, normally 20% above the stalling speed.

4.4 Climb of aircraft with thrust power increasing with speed

The approximation $FV = $ constant has the merit of simplicity and may be nearly correct for a piston engine driving a propeller but, as will be explained in § 5.4.1, most turboprop power plants show an increase of maximum shaft power with forward speed. The propeller efficiency is more likely to increase with speed than to decrease, so that the maximum thrust power usually increases with forward speed and a simple representation of this is an empirical equation of the form

$$FV^n = \text{constant}, \qquad (4.18)$$

where n is a constant which is less than 1. For the small turboprop 'B' represented in Figure 5.19 the value of n would be about 0.5 if the propeller efficiency were independent of forward speed in the range used for climbing. The Equation (4.18) can also be used for civil turbofans and the discussion in § 5.2.2 shows that typical values of n for these engines are in the range 0.2–0.35.

The conditions for maximum rate of climb will now be considered for an aircraft whose thrust varies with speed in accordance with Equation (4.18). For any given height the ratio $f = F/W$ can be written as $f = f_1/v^n$, using an argument similar to that employed in deriving Equation (4.15). Then if $\cos \gamma = 1$ the rate of climb at constant speed is

$$V_c = [vV_e^*/\sigma^{1/2}][f_1/v^n - \tfrac{1}{2}\beta_m(v^2 + v^{-2})]$$
$$= [V_e^*\beta_m/\sigma^{1/2}][\epsilon_1 v^{(1-n)} - \tfrac{1}{2}(v^3 + v^{-1})], \qquad (4.19)$$

where as before $\epsilon_1 = f_1/\beta_m$.

For maximum rate of climb, v must be chosen to make

$$[2\epsilon_1 v^{(1-n)} - v^3 - v^{-1}] \quad \text{maximum}$$

and the condition for this is

$$3v^4 - 2\epsilon_1(1-n)v^{(2-n)} - 1 = 0. \qquad (4.20)$$

Then, with v given by Equation (4.20), the angle of climb is given by

$$\sin \gamma = f - \beta = \tfrac{1}{2}\beta_m[2\epsilon_1/v^n - v^2 - v^{-2}]. \qquad (4.21)$$

As an example, the values of v and γ given by these equations are shown in Figure 4.5 for the case where $n = 0.5$ in Equation (4.18). The

Climb of aircraft with thrust power increasing with speed

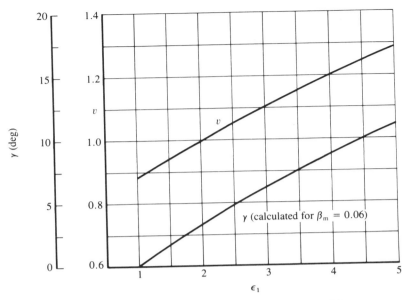

Figure 4.5. Conditions for maximum rate of climb with $FV^{0.5} = $ constant.

values shown for v are valid for any value of β_m but for the calculation of γ it has been assumed that $\beta_m = 0.06$.

When it cannot be assumed that $\cos \gamma = 1$ the equation for γ is obtained by substituting f_1/v^n for f in Equation (4.3) and the equation for the rate of climb becomes

$$V_c = [V_e^* \beta_m / \sigma^{1/2}][\epsilon_1 v^{(1-n)} - \tfrac{1}{2}(v^3 + v^{-1} \cos^2 \gamma)]. \qquad (4.22)$$

For $n = 0.5$ and $\beta_m = 0.06$, these equations give values of v and γ for maximum rate of climb that differ only slightly from those shown in Figure 4.5. As expected, the difference increases with ϵ_1, i.e. with γ, but even for $\epsilon_1 = 5$ the error due to assuming that $\cos \gamma = 1$ is only 0.5% in v and 1% in γ.

Figure 4.5 shows that for typical values of ϵ_1 (in the range 2–4) the change from $n = 1$ to $n = 0.5$ in Equation (4.18) increases the value of v for maximum rate of climb from 0.76 or less to a value in the range 1.0–1.2. The speeds for these higher values of v are likely to be well above the stalling speed in most cases, so that the results given in Figure 4.5 can be regarded as realistic if the variation of thrust with speed can be represented with reasonable accuracy by $FV^{0.5} = $ constant. Results for $FV^n = $ constant with any other value of n can of course be calculated from the Equations (4.19) to (4.22).

4.5 Energy equations

In § 4.1–4.4 it has been assumed that the aircraft climbs at constant speed, i.e. with zero acceleration. In practice the speed does not usually remain constant during a climb and it is then sometimes useful to base the analysis of climbing performance on the changes of energy that occur as the aircraft climbs and accelerates. It will be shown that the equations for zero acceleration derived earlier for rate of climb and for the speed giving maximum rate of climb can easily be interpreted for the general case with non-zero acceleration.

The energy approach to performance problems has been explained by Rutowski (1954). The total energy of an aircraft of mass m is the sum of the potential and kinetic energies

$$E = mgh + \tfrac{1}{2}mV^2 \tag{4.23}$$

and it is useful to define a *specific energy*

$$e = E/m = gh + \tfrac{1}{2}V^2. \tag{4.24}$$

The *energy height* is defined as

$$h_e = E/W = e/g = h + V^2/(2g) \tag{4.25}$$

and this is the height at which the potential energy alone would be equal to E, the current total energy of the aircraft. It should be noted that Rutowski (1954) and some other authors define specific energy as E/W, so that it is the same as energy height, but the definition of Equation (4.24) will be used here.

As fuel is used the mass of the aircraft decreases and correct allowance can be made for this while keeping the equations in a simple form. Since $E = em$ the loss of energy when a mass dm_f of fuel is used is $e\,dm_f$. To explain this more fully, consider an aircraft of mass m from which a mass Δm becomes detached but continues to move at the same speed and height. The energy of the aircraft decreases by $e\,\Delta m$ but this is equal to the energy of the detached part, so that the total energy is unchanged.

Considering a time interval dt in which a mass of fuel dm_f is used, the work done by the engine thrust is $FV\,dt$ and the work done against drag is $DV\,dt$. Thus the total gain of energy is

$$dE = (F - D)V\,dt - e\,dm_f. \tag{4.26}$$

Also, since $E = em$, $dE = m\,de + e\,dm$, where $dm = -dm_f$. Hence

$$dE = m\,de - e\,dm_f \tag{4.27}$$

and Equation (4.26) shows that

$$(F - D)V\,dt = m\,de$$

Energy equations

or

$$de/dt = V(F-D)/m. \tag{4.28}$$

This equation allows correctly for the change of mass as fuel is used, although the same equation can be derived from Equations (4.24) and (4.26) by neglecting the change of mass. In contrast, the equation

$$dE/dt = V(F-D) \tag{4.29}$$

is obtained from Equation (4.26) only by neglecting the change of mass and is therefore only approximate.

Equations (4.24) and (4.28) give

$$de/dt = g\, dh/dt + V\, dV/dt = V(F-D)/m \tag{4.30}$$

or

$$dh/dt = V(F-D)/W - (V/g)\, dV/dt \tag{4.31}$$

and if $dV/dt = 0$ this agrees with Equation (2.10). Equations (4.25) and (4.30) show that the rate of increase of energy height is

$$dh_e/dt = dh/dt + (V/g)\, dV/dt = V(F-D)/W. \tag{4.32}$$

The important quantity $V(F-D)/W$ is known as the *specific excess power* P_{se} and if V_c' is used to denote dh_e/dt, with $V_c = dh/dt$ as before, Equation (4.32) shows that in general $P_{se} = V_c'$, while in the special case of zero acceleration $P_{se} = V_c$. In § 4.2–4.4 all the equations giving the rate of climb V_c with zero acceleration are based on the equation $V_c = P_{se}$, whether or not $\cos \gamma$ is assumed to be equal to 1. Thus all these equations can also be used to obtain V_c' in the general case with non-zero acceleration. Moreover all the results in § 4.2–4.4 giving the speed for maximum V_c also give the speed for maximum V_c' in the general case, but it will be explained in § 4.7 that these speeds refer to the maximum V_c' at a constant height h and not at a constant energy height h_e. With this reservation the values of v shown in Figures 4.3 and 4.5 give the conditions for maximum V_c' in the general case, although the values of γ given in these figures and the equations such as (4.12) giving $\sin \gamma$ can be used directly only when the acceleration is zero, because in the general case these equations give the value of

$$V_c'/V = V_c/V + (1/g)\, dV/dt$$

and this quantity is not of much interest. Nevertheless it will be shown in § 4.6 that when there is a known acceleration the angle of climb can easily be found.

These conclusions from the energy equations are useful in two ways. First, it will be shown in § 4.6 that Equation (4.31) can be used to correct a rate of climb calculated for zero acceleration in order to allow for the effect of acceleration. Second, since a climbing aircraft is usually required to achieve a change of speed as well as a change of height, it is often useful to base the whole calculation of climb performance on change of energy height and this will be discussed in § 4.7.

4.6 Accelerated climbs

In an accelerated climb

$$dV/dt = (dh/dt)\,dV/dh = V_c\,dV/dh \qquad (4.33)$$

and Equation (4.31) shows that

$$V_c + (V/g)\,dV/dt = V(F-D)/W = P_{se} \qquad (4.34)$$

or

$$V_c[1 + (V/g)\,dV/dh] = P_{se}.$$

Thus

$$V_c = P_{se}/A, \qquad (4.35)$$

where

$$A = 1 + (V/g)\,dV/dh \qquad (4.36)$$

and A is known as the *acceleration factor*, because it can be used to correct a rate of climb calculated for constant speed to allow for the effect of acceleration. Since $\sin\gamma = V_c/V$, the factor can also be used to correct a value of $\sin\gamma$ calculated on the assumption of constant speed and since $P_{se} = dh_e/dt = V_c'$ Equation (4.35) shows that the acceleration factor is equal to the ratio V_c'/V_c. The acceleration factor can also be found very simply by inserting an inertia term in Equation (2.2), giving

$$F = m\,dV/dt + D + W\sin\gamma. \qquad (4.37)$$

Then

$$V_c = V\sin\gamma = V(F-D)/W - (V/g)\,dV/dt$$

as in Equation (4.34). The acceleration factor is also known sometimes as the kinetic energy correction factor or KE factor, since the correction can be regarded as an allowance for the change of kinetic energy while the aircraft accelerates.

As an example, to illustrate the effect of acceleration on the rate of climb, a climb at constant EAS will be considered. This procedure is often convenient, but it should be noted that Figures 4.3 and 4.5 show that the EAS for maximum V_c or V_c' decreases with increasing height,

Accelerated climbs

as the thrust decreases. For this reason, when a climb is made at constant V_e the chosen value of V_e is often adjusted downward at one or more specified heights, so that V_e remains close to the value giving maximum V_c'.

For constant V_e the term $(A - 1)$ can be expressed as

$$(V/g)\,dV/dh = (V/g)(dV/d\sigma)\,d\sigma/dh$$

and using the equation

$$V = V_e \sigma^{-1/2}$$

it becomes

$$(V/g)\,dV/dh = -\tfrac{1}{2}V^2(d\sigma/dh)/(g\sigma). \tag{4.38}$$

For $h \leq 11$ km in the ISA Equations (1.2) and (1.6) show that

$$\sigma = \rho/\rho_0 = (1 - Lh/T_0)^{(g/(RL)-1)}$$

and hence

$$(1/\sigma)\,d\sigma/dh = -(g - RL)/(RT) = -\gamma_a(g - RL)/a^2, \tag{4.39}$$

where γ_a is the ratio of specific heat capacities for air. With $\gamma_a = 1.4$ and the values of R and L given in Appendix 2, Equations (4.38) and (4.39) show that for this climb at constant EAS

$$(V/g)\,dV/dh = \tfrac{1}{2}\gamma_a M^2(1 - RL/g) = 0.567\,M^2, \tag{4.40}$$

where M is the Mach number V/a and a is the velocity of sound.

For 11 km $\leq h \leq 20$ km in the ISA the temperature has the constant value T_1 and it can be shown from Equation (1.8) that

$$(1/\sigma)\,d\sigma/dh = -g/(RT_1)$$

and so Equation (4.40) must be replaced by

$$(V/g)\,dV/dh = \tfrac{1}{2}\gamma_a M^2 = 0.7\,M^2, \tag{4.41}$$

again valid for a climb at constant EAS.

As mentioned earlier the Mach number is sometimes kept constant while climbing instead of the EAS, especially when ϵ is large and the climbing speed is governed mainly by the need to keep the Mach number below the value M_D at which the steep rise in C_D starts. For $h < 11$ km in the ISA the velocity of sound decreases as the height increases and the acceleration must be negative if M is to be kept constant. Putting $V = M(\gamma_a RT)^{1/2}$ and $T = T_0 - Lh$ as given by Equation (1.2) it can be shown that with $\gamma_a = 1.4$ and the values of R and L given in Appendix 2, the term $(A - 1)$ becomes

$$(V/g)\,dV/dh = -\tfrac{1}{2}M^2 RL\gamma_a/g = -0.133\,M^2. \tag{4.42}$$

When $M = 2.74$ the acceleration factor falls to zero and since the factor is equal to P_{se}/V_c this means that the aircraft can climb at any angle with $P_{se} = 0$, i.e. with $F = D$, because for this climb at constant Mach number the gain of potential energy is exactly balanced by the loss of kinetic energy. When $M > 2.74$ the acceleration factor becomes negative and the aircraft can climb at constant M with $P_{se} < 0$, i.e. with $F < D$, implying a loss of energy height as the climb proceeds. Clearly these conclusions are valid only for the atmospheric conditions which have been specified, i.e. for the ISA with $h < 11$ km.

EXAMPLE 4.3. EFFECT OF ACCELERATION ON CLIMB PERFORMANCE

An aircraft with $K_1 = 0.018$ and $K_2 = 0.045$ climbs at a constant EAS of 120 m/s. At a height of 5 km, $f = 0.2$ and the wing loading is 4 kN/m². Calculate the angle of climb and rate of climb (a) without and (b) with allowance for acceleration.

For the calculation of angle of climb and rate of climb in case (a) the value of β is required and this depends on β_m and v. Thus the preliminary calculations are

$$\beta_m = 2(K_1 K_2)^{1/2} = 0.0569,$$
$$V_e^* = (2w/\rho_0)^{1/2}(K_2/K_1)^{1/4} = 101.6 \text{ m/s},$$
$$v = V_e/V_e^* = 1.181,$$
$$\beta = \tfrac{1}{2}\beta_m(v^2 + v^{-2}) = 0.0601$$

and from Appendix 2 the velocity of sound at sea level is $a_0 = 340$ m/s, while at $h = 5$ km the density and pressure ratios are $\sigma = 0.6009$ and $p/p_0 = 0.5331$.

(a) Without allowance for acceleration,

$$\sin \gamma = f - \beta = 0.140 \quad (\text{assuming } \cos \gamma = 1)$$

from which $\gamma = 8.04°$ and the rate of climb follows as

$$V_c = (V_e/\sigma^{1/2}) \sin \gamma = 21.7 \text{ m/s}.$$

(b) With allowance for acceleration, Equation (2.8) shows that

$$\text{Mach number} = M = V/a = (V_e/a_0)(p_0/p)^{1/2} = 0.483$$

and thus

$$(A - 1) = (V/g) \, dV/dh = 0.567 M^2 = 0.132$$

so that

$$A = 1.132.$$

Then

$$\sin \gamma = 0.140/1.132 = 0.1237,$$

Accelerated climbs

from which the corrected angle of climb is found to be

$$\gamma = 7.10°$$

and the corrected rate of climb is $V_c = 21.7/1.132 = 19.2$ m/s.

In general, the speed for maximum rate of climb is also affected by acceleration but this speed depends on what is kept constant as the speed is varied. If the acceleration factor A remains constant Equation (4.35) shows that the speed for maximum V_c is the same as the speed for maximum P_{se} and is independent of acceleration, but with V_c nearly constant in the region of the maximum, constant A implies an acceleration that is inversely proportional to speed and this is not realistic. It is more realistic to assume that $(1/V)\,dV/dh$ remains constant as in a climb at constant EAS, so that the acceleration is directly proportional to speed. With this assumption the effect of acceleration on the speed for maximum rate of climb will now be examined, for an aircraft with thrust independent of speed at a given height.

Writing $V_h = (1/V)\,dV/dh$ and with $\cos\gamma = 1$, Equations (4.7), (4.35) and (4.36) show that the rate of climb is

$$V_c = \frac{V^*\beta_m[v\epsilon - \tfrac{1}{2}(v^3 + v^{-1})]}{1 + V_h(vV^*)^2/g}, \qquad (4.43)$$

where $V^* = V_e^*/\sigma^{1/2}$, the true air speed for minimum β. For constant V_h it can be shown that the condition for V_c to be maximum is

$$(A+2)v^4 + 2\epsilon(A-2)v^2 - (3A-2) = 0 \qquad (4.44)$$

and this gives

$$v^2 = \{\epsilon(2-A) + [\epsilon^2(2-A)^2 + (3A-2)(A+2)]^{1/2}\}/(A+2), \qquad (4.45)$$

but it should be noted that although this equation gives v in terms of ϵ and A, the maximum V_c that is obtained is at constant V_h (and ϵ), not at constant A. Values of v given by Equation (4.45) are shown in Figure 4.6 for a range of values of A and ϵ, where the curve for $A=1$ is of course the same as the unbroken curve for v in Figure 4.3. The speed ratio v for maximum rate of climb is reduced when the acceleration is positive $(A>1)$ because with constant V_h a decrease of speed gives a reduction of A and tends to increase V_c. When $\epsilon = 1$ the optimum value of v is 1.0 and is independent of A, because both the rate of climb and the acceleration are then zero.

4.7 Climb performance in terms of energy height

Figure 4.7 shows the relations between h_e, V and h as given by Equation (4.25). This shows that in a typical subsonic cruise at $M = 0.8$ and $h = 11$ km the energy height exceeds the true height by about 2.8 km, indicating that the kinetic energy is about 20% of the total energy, whereas if the speed is doubled to $M = 1.6$ the difference $(h_e - h)$ becomes about 11 km and at the same height the kinetic energy is 50% of the total.

The usual requirement for a climb is that the aircraft should reach specified values of both the height and the speed, either in a minimum time or with minimum use of fuel or in accordance with some other requirement such as minimum 'lost range' as will be discussed in Chapter 7. In order to achieve the required objective the speed of the aircraft should be continually adjusted to the optimum value at every stage of the climb and there is a need for calculation of this optimum

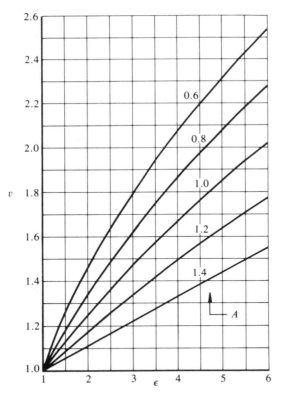

Figure 4.6. Effect of acceleration factor A on speed for maximum rate of climb, with constant thrust and constant $(1/V) \, dV/dh$.

Climb performance in terms of energy height

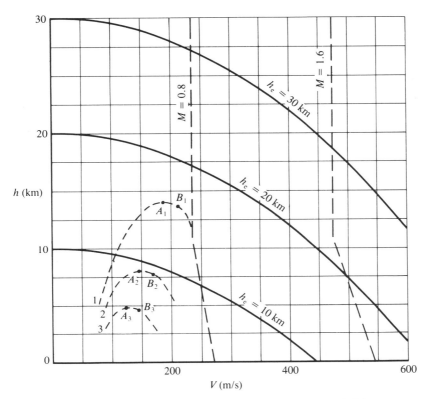

Figure 4.7. Relations between energy height, speed and true height.
----- Lines of constant P_{se} (P_{se} increasing from 1 to 3).

speed. The true optimum speed will not be found if only the required gain of height is considered and the required final speed is ignored, for then the climb procedure will in general give the correct height but the wrong speed at the end of the climb. For any high speed aircraft and particularly for a supersonic one it is more satisfactory to specify both the initial and final states in terms of energy height and consider the optimum procedure for gaining energy height. The total climb is then considered in three stages, as follows:

(1) a transition from the initial height and speed to the optimum height and speed for the start of stage (2),
(2) an 'energy climb' in which the speed is continually adjusted to the optimum value for gain of energy height,
(3) a transition to the required final height and speed.

Stages (1) and (3) usually take only a short time and are planned to simplify the task of the pilot, e.g. a level acceleration or a climb at constant M may be used. If the pilot embarks on such an exercise immediately after take-off, stage (1) is usually an accelerating climb at a small angle. For a high speed military aircraft stage (3) may take the form of a quick interchange between kinetic and potential energies, i.e. a dive or a zoom. For a transport aircraft such manoeuvres are not acceptable and stage (3) may often start before the required final energy height has been reached, with a gradual acceleration or climb to achieve the required final height and speed.

The optimum procedure for the energy climb of stage (2) will now be considered. The time taken to change between the initial and final energy heights h_{e1} and h_{e2} is

$$t = \int_{h_{e1}}^{h_{e2}} \frac{1}{dh_e/dt} \, dh_e = \int_{h_{e1}}^{h_{e2}} \frac{1}{P_{se}} \, dh_e. \tag{4.46}$$

Thus for minimum time the requirement is that P_{se} should be maximum at every stage of the climb, but as pointed out by Rutowski (1954) the integral in Equation (4.46) shows that the speed at each level of h_e should be adjusted to give maximum P_{se} at constant h_e and not at constant h. The distinction between the speeds for maximum P_{se} at constant h_e and at constant h can be seen by referring to Figure 4.7, where the broken curves 1, 2 and 3 are lines of constant P_{se}, with the value of P_{se} increasing from 1 to 3. The speeds for maximum P_{se} at constant h are given by the points A_1, A_2, A_3, where the tangent to the curve is a line of constant h, whereas for constant h_e the speeds for maximum P_{se} are given by the points B_1, B_2, B_3, where the tangent to the curve is also a tangent to a curve of constant h_e. Thus the speed for the required condition of maximum P_{se} at constant h_e is greater than that giving the maximum at constant h. The difference between the two speeds is fairly small for subsonic speeds but it becomes larger as the speed increases in the supersonic range and the negative slopes of the lines of constant h_e become larger.

The results given in § 4.2–4.4 relating to the speed for maximum V_c (with zero acceleration) give the conditions for maximum P_{se} at a given true height h and are therefore only approximately correct for obtaining the minimum values of the time given by Equation (4.46). (It should also be noted that in § 4.2–4.4 it has been assumed that the simple parabolic drag law is valid, with constant values of K_1 and K_2, and this assumption is correct only for $M < M_D$ as discussed in Chapters 2 and 3.)

When a more exact calculation is required to find the minimum time to climb between two given values of h_e a graphical method may be used to give both the required speed at each level of h_e and the total time. There are several ways of doing this, one of which is based on a diagram like Figure 4.7. A family of lines of constant P_{se} is constructed and on each curve the point B is marked where the tangent to the curve is also a tangent to a line of constant h_e. These points give both P_{se} and the optimum speed as functions of either h or h_e. Hence the optimum climb procedure can be specified and the minimum time can be found by numerical evaluation of the integral in Equation (4.46).

If Q is the mass of fuel used per unit time, the mass of fuel used in an energy climb from h_{e1} to h_{e2} is

$$m_f = \int_{h_{e1}}^{h_{e2}} \frac{dm_f/dt}{dh_e/dt} dh_e = \int_{h_{e1}}^{h_{e2}} \frac{Q}{P_{se}} dh_e \qquad (4.47)$$

and if this is required to be minimum the speed should be chosen at each level of h_e so that P_{se}/Q has the maximum value at constant h_e. The required climb procedure and the total mass of fuel used can be found by a graphical method similar to that used to find the minimum time, but in this case lines of constant P_{se}/Q should be plotted.

In § 7.10 the effect of the initial climb on the range obtainable with a given mass of fuel will be discussed and it will be shown that the optimum climbing speed is then greater than the value giving minimum time for the specified gain of h_e.

4.8 Maximum angle of climb

The speed required to give maximum angle of climb will now be considered. If it is assumed that $\cos \gamma = 1$ and there is no acceleration, Equation (2.3) shows that $\sin \gamma = f - \beta$. Then if f is independent of speed, for any given height, the maximum angle of climb is given by

$$\sin \gamma = f - \beta_m = \beta_m(\epsilon - 1) \qquad (4.48)$$

and this is obtained at $v = 1$. The full line in Figure 4.8 shows the maximum value of γ as a function of ϵ, for $\beta_m = 0.06$.

If it cannot be assumed that $\cos \gamma = 1$ Equation (4.13) must be used instead of (4.48). The effect of this on the maximum angle of climb is very small, as shown by the broken line for γ in Figure 4.8, but the value of v required to give maximum γ is reduced when ϵ becomes large, as shown by the broken line for v in Figure 4.8. The reason for this effect of the $\cos \gamma$ term on v can be seen by inspection of Equation (4.13). The $\cos^2 \gamma$ term in that equation reduces the importance of the $(\frac{1}{2}v^{-2})$ component of β/β_m that is shown in Figure 3.5 and this reduces the value of v for minimum β. Figure 4.8 shows that the effect becomes significant when the maximum angle of climb exceeds about $20°$. For $\epsilon = 1/\beta_m$ the calculated value of v for maximum angle would fall to zero, because then $F = W$ and the balance of forces which

70 *Climbing performance*

served as the foundation of these analyses would be satisfied by the condition of a vertical 'climb' at zero speed, for which both the lift and the drag would also be zero, i.e. the aircraft would be static and supported only by its thrust.

When the thrust power FV is constant the angle of climb with zero acceleration is given by Equation (4.15) if $\cos \gamma = 1$. This equation shows that the value of v for maximum angle of climb is given by

$$v^4 + \epsilon_1 v = 1, \tag{4.49}$$

where $\epsilon_1 = f_1/\beta_m$. This equation gives $v = 0.76$ for $\epsilon_1 = 0.878$, the condition for flight at the absolute ceiling (with zero rate of climb). As ϵ_1 increases above this value the angle of climb increases and the value of v given by the equation falls to levels that are quite unrealistic, e.g. $v = 0.6$ for $\beta_m = 0.06$ when the climb angle is only 3°. It was noted earlier that for angles of climb greater than about 15° the speed used in practice to give maximum rate of climb would be the lowest that is compatible with safety. The results obtained from Equation (4.49) now show that this same speed must be used to give the maximum practicable angle of climb, even when the angle is quite small.

In § 4.4 the conditions for maximum rate of climb were considered for the thrust–speed relation given by Equation (4.18). If this equation

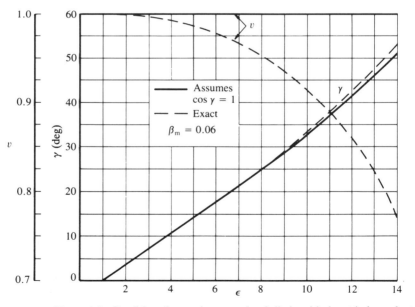

Figure 4.8. Conditions for maximum angle of climb, with thrust independent of speed.

Maximum angle of climb

is used to determine the speed for maximum angle of climb, it is found that the values of v obtained are too low to be realistic unless n is less than about 0.3 and it will be shown in Chapter 5 that n is likely to be about 0.2–0.4 for a civil turbofan, 0.5 for a turboprop and 0.6–0.7 for a propfan. Thus for a civil turbofan the values of v calculated for maximum angle of climb may be realistic, but for a turboprop or propfan the speed used in practice for maximum angle of climb will be determined by the need to avoid stalling and maintain safety.

It may be useful at this stage to refer again to Figures 2.9 and 2.12 which show V_{ec} as the EAS for maximum angle of climb, i.e. maximum separation between the f and β curves. For the cases shown in these figures the ratio f decreases only slowly with increasing speed and as a consequence V_{ec} is not much below V_e^*. As explained in the discussion of Figure 2.12 in §2.7 there is speed instability if $V_e < V_{ec}$ and the speed for maximum angle of climb ($V_e = V_{ec}$) gives neutral speed stability. Since $d\gamma/dV_e$ is zero at the speed for maximum γ a moderate increase of V_e above V_{ec} will cause only a small loss of γ and will have the two advantages of ensuring positive speed stability and promoting safety by increasing the margin of speed above the stalling speed.

In any case for which there are difficulties in representing f and β by equations the angle of climb and its maximum value can be found by the graphical method explained in Chapter 2, provided that numerical data are available giving f and β as functions of speed and height.

When the aircraft is accelerating during the climb the value of $\sin \gamma$ calculated for zero acceleration can be corrected by introducing the acceleration factor as explained in §4.6. The speed for maximum angle of climb is unaffected by acceleration if A remains constant as the speed changes, but as discussed earlier in considering the maximum rate of climb an assumption of constant A is not realistic and a more appropriate assumption is that $V_h = (1/V)\, dV/dh$ remains constant as the speed changes. For an aircraft with thrust independent of speed at a given height Equation (4.43) shows that the angle of climb is given by

$$\sin \gamma = \frac{V_c}{V} = \frac{\beta_m[\epsilon - \tfrac{1}{2}(v^2 + v^{-2})]}{1 + V_h(vV^*)^2/g} \qquad (4.50)$$

and for constant V_h it can be shown that the condition for γ to be maximum is

$$v^4 + 2\epsilon(A-1)v^2 - (2A-1) = 0 \qquad (4.51)$$

and this gives

$$v^2 = \epsilon(1-A) + [\epsilon^2(1-A)^2 + (2A-1)]^{1/2}. \qquad (4.52)$$

72 Climbing performance

Values of v given by Equation (4.52) are shown in Figure 4.9, where the simple result $v = 1$ appears for zero acceleration ($A = 1$). As found earlier for the maximum rate of climb the optimum speed ratio is reduced when the acceleration is positive ($A > 1$) and increased when it is negative, but it should be noted that values of v that are much below 1 may be unrealistic because the corresponding speeds may be close to the stalling speed.

4.9 Rate of climb in a non-standard atmosphere

When an aircraft is climbing in an atmosphere which has properties different from those of the ISA a correction must be applied to the observed rate of change of altimeter reading in order to obtain the true rate of climb.

A correctly calibrated altimeter reads directly the pressure height h_s, the height in the ISA at which the air pressure would be equal to the measured pressure p. If ρ_s and T_s are the density and temperature in

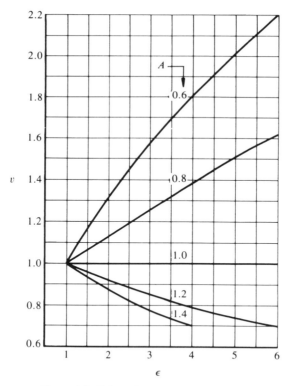

Figure 4.9. Effect of acceleration factor A on speed for maximum angle of climb, with constant thrust and constant $(1/V)\,dV/dh$.

Rate of climb in a non-standard atmosphere

the ISA at this height, the apparent rate of climb as given by the rate of change of altimeter reading is

$$dh_s/dt = (dh_s/dp)\, dp/dt = -(dp/dt)/(g\rho_s). \tag{4.53}$$

If h is the true height above sea level and ρ and T are the true density and temperature, the true rate of climb is

$$dh/dt = (dh/dp)\, dp/dt = -(dp/dt)/(g\rho)$$
$$= (\rho_s/\rho)\, dh_s/dt = (T/T_s)\, dh_s/dt. \tag{4.54}$$

Thus the true rate of climb dh/dt can be obtained from the apparent rate of climb dh_s/dt, simply by measuring the true temperature T and using the temperature ratio T/T_s as a correction factor.

5
Power plants

The thrust F produced by the engines is of great importance in almost every phase of flight because it counteracts the drag and enables the aircraft to climb if required. The *maximum available* thrust F_m depends on the height and speed of the aircraft and is limited by the approved 'rating' for the appropriate phase of the flight. The three ratings that are important in relation to aircraft performance calculations are those specified for take-off, climb and cruise, and the rated thrust for each of these is the maximum available. In any phase of flight the thrust can of course be reduced by the pilot below the rated value, usually by moving a single control lever which is commonly known as a 'throttle' lever, even though it may act on a complex engine control system.

For almost all aspects of aircraft performance calculation it is necessary to know how F_m varies with the speed and height of the aircraft. In addition, for calculations of range, endurance and operating cost, a knowledge is required of the rate of consumption of fuel and the way in which this varies with flight speed, height and engine throttle setting. In this chapter the principles governing these variations will be discussed and approximate equations will be introduced for representing the variations in calculations of aircraft performance. For this purpose the rate of consumption of fuel will be expressed as the ratio of the rate of consumption to either the thrust or the shaft power of the engine.

All the types of power plant that are used for aircraft propulsion at subsonic and moderate supersonic speeds are air-breathing systems which depend for the generation of thrust on the acceleration of a stream of air drawn into the system. These include such apparently different systems as turbojet engines and propellers driven by piston engines. The rockets that are used for high supersonic speeds and space flight are not air-breathing and will not be considered here.

A wide range of air-breathing power plants is available, in most of which a proportion of the accelerated air stream participates in the

Efficiency of thrust generation

process of power generation by burning of fuel. Nevertheless it is instructive to consider the efficiency of thrust generation by accelerating an air stream, separately from the efficiency of the thermodynamic process by which mechanical power for accelerating the air stream is generated by burning fuel, since the former can be examined by simple mechanics whereas the latter requires a study of the thermodynamics of the engine.

It is possible by means of cycle calculations to analyse in detail the flow through an engine and predict its performance, but calculations of this kind will not be considered here. In studies of aircraft performance the only important quantities relating to the power plant (apart from its weight and the drag of the installation) are the maximum available thrust and the rate of consumption of fuel and attention will be concentrated on these two quantities.

5.1 Efficiency of thrust generation

Figure 5.1 shows the essential features of an air-breathing propulsive system, in which air is drawn in at velocity V and ejected at the increased velocity V_J after energy has been added in the zone E. The bounding streamlines AB and CD are drawn here in a form that is representative of an unducted propeller, but the equations to be derived do not depend on the form of these streamlines and are valid for all forms of air-breathing propulsive system. In reality the increase of velocity is usually not uniform across the outlet but for the purpose of this simple analysis it is assumed that the velocities V and V_J are uniform across the sections AC and BD. It is also assumed that AC and BD are respectively so far upstream and downstream that the pressure at each of these stations is equal to the pressure in the undisturbed air. The velocity V at inlet to the system is then equal to the flight speed and consideration of the change of momentum shows that the thrust is

$$F = \dot{m}(V_J - V), \tag{5.1}$$

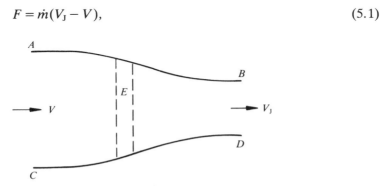

Figure 5.1. Propulsive system.

where $\dot{m}$ is the mass of air flowing through the system per unit time. (In the case of a turbojet or turbofan engine the small addition to the mass flow rate due to injection of fuel into the engine is neglected.)

The power required ideally to accelerate the air stream from the velocity V to V_J is $\frac{1}{2}\dot{m}(V_J^2 - V^2)$ and the *propulsive efficiency* η_p is defined as the ratio of the useful thrust power FV to this ideal power. Thus

$$\eta_p = 2V(V_J - V)/(V_J^2 - V^2) = 2/(1 + V_J/V). \quad (5.2)$$

The velocities V and V_J are the inlet and outlet velocities relative to the aircraft and the system leaves behind the aircraft a stream having a velocity $(V_J - V)$ relative to the undisturbed air. The kinetic energy added to this stream per unit time is $\frac{1}{2}\dot{m}(V_J - V)^2$ and this represents 'wasted' power. The power required ideally to accelerate the air is the sum of this wasted power and the thrust power FV. Thus an alternative and exactly equivalent definition of propulsive efficiency is

$$\eta_p = FV/[FV + \tfrac{1}{2}\dot{m}(V_J - V)^2]$$

and this is equal to $2/(1 + V_J/V)$ as found earlier.

For generation of thrust V_J/V must be greater than 1, but Equation (5.2) shows that it should not be much greater if high propulsive efficiency is required. Equation (5.1) may be written as

$$F = \dot{m}V(V_J/V - 1)$$

and this shows that if V_J/V is kept down to a value that is only a little greater than 1, to ensure high propulsive efficiency, it is necessary to have a large value of $\dot{m}V$ in order to generate a given thrust. Thus a high mass flow rate is required unless the flight speed is high. A quantity that is often used to describe the characteristics of a turbojet or turbofan engine is the *specific thrust* F_s, defined as $F/\dot{m}$, and for the simple model considered here

$$F_s = V_J - V$$

so that

$$\eta_p = 1/(1 + \tfrac{1}{2}F_s/V). \quad (5.3)$$

The requirement given earlier that for high η_p the product $\dot{m}V$ should be large for a given thrust is equivalent to the requirement shown by Equation (5.3) that F_s/V should be small.

The overall efficiency of the power plant depends not only on η_p but also on the efficiency of the thermodynamic process in the engine. For a turbojet or turbofan engine the *thermal efficiency* η_{th} is defined as the ratio of the rate of addition of kinetic energy to the rate of energy generation by combustion of fuel. As implied by this definition the primary function of the engine is to raise the kinetic energy of the stream of air passing through it. There are necessarily some aerodynamic and thermodynamic losses, so that not all of the energy released by combustion of fuel is available for accelerating the air and the thermal efficiency η_{th} is less than 1. The overall efficiency is simply defined as the product of the two separate efficiencies and is

$$\eta_o = \eta_p \eta_{th}. \tag{5.4}$$

The rate of consumption of fuel in a power plant is usually expressed in terms of a *specific fuel consumption* (sfc). For turbojet and turbofan engines this is defined as $c = Q/F$, where Q is the mass of fuel used per unit time and F is the thrust. This definition is convenient because for turbojets and for turbofans with low by-pass ratio (to be defined in § 5.2) the variation of c with aircraft speed is relatively small within the range of speeds used for cruising. For turboprop power plants a different definition of sfc is often used, because in this case the rate of consumption of fuel is roughly proportional to the shaft power that is available for driving the propeller. For given values of the shaft power and propeller efficiency the thrust is inversely proportional to the flight speed, so that the quantity c as defined above would be roughly proportional to flight speed. It is more useful to define the sfc for these engines so that it is related to power instead of thrust and therefore does not vary much with flight speed. The definition that is commonly used in performance calculations is $c' = Q/P_e$, where P_e is the equivalent shaft power. The significance of the word 'equivalent' is that allowance can be made for the effects of direct jet thrust as will be explained in § 5.4.

5.2 Turbojet and turbofan engines

Figure 5.2 shows in diagrammatic form the essential features of a turbojet engine. Air from the intake passes through the multi-stage compressor C and then through the combustion chamber CT where fuel is burned and there is a large rise of temperature. The hot gas at high pressure then passes through the turbine T where

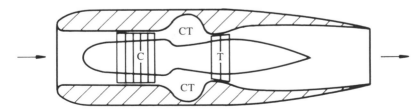

Figure 5.2. Turbojet engine.

sufficient power is extracted from the gas stream to drive the compressor. The gas leaving the turbine passes through a propelling nozzle where its velocity increases and there is a corresponding decrease of pressure. In some cases the nozzle may be choked, so that the final expansion to atmospheric pressure can occur only after the gas leaves the nozzle exit, with the gas accelerating to supersonic velocity.

A more usual form of engine is the turbofan, shown diagrammatically in Figure 5.3. In this type of engine the air from the intake passes first through the fan F and part of the flow then passes through an annular duct outside the hot core engine directly to an annular outlet nozzle, without passing through a combustion chamber or turbine. The remaining part of the flow passes through the compressor C, the combustion chamber CT and the turbine T as in the turbojet engine. The turbine is designed to extract sufficient power from the gas stream to drive both the compressor C and the fan F. The diagrammatic form of Figures 5.2 and 5.3 should be emphasised; in particular the detailed arrangements of the combustion chambers are not shown and the gas ducting is not drawn to scale. More details are shown in Figure 5.4, which is a cutaway view of a large modern turbofan engine, the Rolls–Royce RB211-524G, for use in subsonic civil aircraft. For further information on turbojets and turbofans reference may be made to Cohen, Rogers & Saravanamuttoo (1987).

The part of the flow in a turbofan engine that does not pass through the hot core is known as the *by-pass* flow and early engines of this type were known as by-pass jet engines, but the term turbofan is now more usual. The *by-pass ratio* λ is the ratio of the mass flow rate in the by-pass to that in the hot core engine (combustion chamber and turbine), often called the 'gas generator'. Referring to Figure 5.3, the by-pass exit velocity V_{J2} is normally much lower than the core jet velocity V_{J1}. As the gases pass downstream the two gas streams become mixed, giving a jet of mean velocity V_J which is lower than V_{J1} and decreases as the by-pass ratio λ is increased. In some engines,

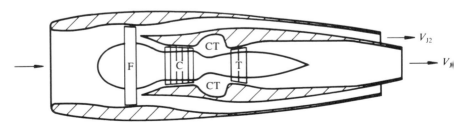

Figure 5.3. Turbofan engine.

Turbojet and turbofan engines

Figure 5.4. Rolls–Royce RB211-524G civil turbofan.

known as mixed-stream turbofans, the ducts are designed so that mixing occurs before the nozzle exit is reached.

In comparison with a turbojet the reduced value of the mean jet velocity V_J in a turbofan gives two important advantages. First, it gives an increase of propulsive efficiency as shown by Equation (5.2) and this leads to a reduction of sfc. Second, the reduction of jet velocity gives a large reduction of noise. Smith (1989) has reviewed the available data on the noise generated by mixing of the jet with the external air and refers to the theory given by Lighthill (1952), showing that the acoustic power should be proportional to the 8th power of the jet velocity, so that if the jet velocity is halved the acoustic power is reduced by a factor of about 250, i.e. by 24 decibels. Smith (1989) shows that this 8th power law is supported by experiments with cold air jets, but measurements on engines with relatively low jet velocities give a more gradual increase of noise level with velocity, because of effects of temperature differences and the presence of sound sources other than jet mixing. Nevertheless, for the jet velocities that are relevant, reduction of the mean velocity by the introduction of by-pass flow does give a large reduction of noise, although the mixing processes for the by-pass and core flows and the external air stream are complex and depend on the design of the engine ducts and nozzles.

Because of these advantages the turbofan has entirely superseded the turbojet for subsonic flight and in Chapters 6–9 turbojets will not be considered. For supersonic flight a high jet velocity V_j is required for the generation of thrust, since Equation (5.1) shows that V_j must be greater than V. There is also a need for high specific thrust in order to keep the frontal area of the engine small and thus reduce the drag. For these reasons turbojets may sometimes be used for supersonic flight, as in Concorde with its Olympus 593 engines, but for supersonic combat aircraft the usual choice is a mixed stream turbofan with a low by-pass ratio.

The by-pass ratio λ of a turbofan varies to some extent with the operating state of the engine. The early by-pass engines had values of λ less than 1, but for modern turbofans in subsonic aircraft λ is usually between about 3 and 6 and the term 'civil turbofan' will be used to denote these engines. For military aircraft required to operate at both subsonic and supersonic speeds the optimum value of λ is considerably lower and values between 0 and 1 are commonly used, depending on the required compromise between subsonic and supersonic fuel economy. The term 'military turbofan' will be used for these engines.

For either a turbojet or a turbofan engine the term $\dot{m}V$ in Equation (5.1) is known as the *intake momentum drag,* because it is equal to the momentum flux in the air stream approaching the intake and represents a force that opposes the thrust. The thrust given by the term $\dot{m}V_j$ in the equation is known as the *gross thrust* and the difference between this and the intake momentum drag is the *net thrust F,* which is the thrust required for performance calculations and is simply called the thrust throughout this book.

With either a turbojet or a turbofan engine the thrust can be greatly increased by *reheat,* i.e. by burning additional fuel in an *afterburner* in the jet pipe, upstream of the final exit nozzle. There are severe penalties in high specific fuel consumption and noise but reheat is often used for short periods in supersonic aircraft, e.g. for take-off, for acceleration from subsonic to supersonic speeds, for combat manoeuvres and sometimes for all supersonic flight. The performance of engines with reheat will be considered in § 5.2.4, but in other sections of this chapter and in Chapters 6–9 it will be assumed that reheat is not used.

5.2.1 NON-DIMENSIONAL RELATIONS

As shown by Cohen *et al.* (1987) the characteristics of a turbojet or turbofan engine may be expressed as functional relations between non-dimensional quantities. If effects of varying Reynolds number within the engine are neglected and if the same fuel is always used, elimination of variables which are constant for a given engine of

fixed geometry then leads to the following functional relations:
$$F/\delta = f_1(N/\theta^{1/2}, M) \tag{5.5}$$
and
$$Q/(\delta\theta^{1/2}) = f_2(N/\theta^{1/2}, M), \tag{5.6}$$
where N is the rotational speed of the engine, $\theta = T/T_0$, $\delta = p/p_0$ and Q is the mass of fuel burned per unit time. In the case of an engine with two or more shafts, N is usually defined as the speed of the fan or low pressure compressor. The relations are then correct only if the ratios of this speed to the other shaft speeds remain constant; if this condition is not satisfied the speed ratios must be included as independent variables in addition to $N/\theta^{1/2}$ and M. The relations (5.5) and (5.6) show that the sfc may be expressed as
$$c = Q/F = \theta^{1/2}f_3(N/\theta^{1/2}, M) \tag{5.7}$$
and all these relations are useful in two ways. First, they provide a useful means of plotting data on thrust and sfc as shown by the typical curves in Figures 5.5 and 5.6, but it should be emphasised that the

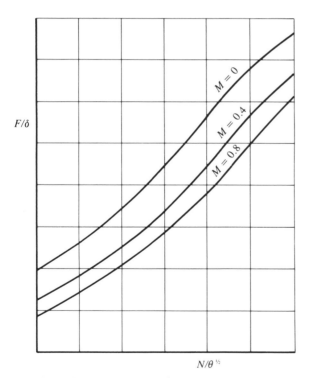

Figure 5.5. Thrust function (5.5) for a turbofan.

curves depend on the characteristics of the engine and may not always have the form shown in these figures. If complete plots of this kind are available for an engine the thrust and sfc can be found for any specified combination of engine speed, flight Mach number and atmospheric temperature and pressure. Second, even if the exact forms of the functions f_1 and f_3 are not known some important deductions can be made from the functional relations. In particular the relations show that for specified values of the engine speed and operating height (in the ISA) both F and c depend only on the flight Mach number M. Also for *any* height in the ISA in the range 11–20 km, where θ is constant, F/δ and c depend only on M for a given engine speed. In general the effect of increasing air temperature is important because it is equivalent to a reduction of engine speed and causes a loss of thrust as shown in Figure 5.5.

5.2.2 MAXIMUM THRUST OF CIVIL TURBOFANS

The maximum thrust F_m is obtained at the appropriate rating as mentioned earlier, for take-off, climb or cruise, and for performance calculations a knowledge of the variation of F_m with flight speed and height is required. Considering first the effect of varying flight speed at a constant height, the rotational speed N for maximum thrust is fixed by the rating and the relation (5.5) shows that the thrust depends only on the Mach number (or on the true air speed, since the velocity of sound is constant at a fixed height). As the Mach number

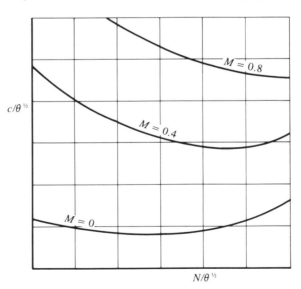

Figure 5.6. sfc function (5.7) for a turbofan.

Turbojet and turbofan engines

or speed increases from a low value at a fixed height, with constant N, there are two main effects on the thrust F. First, there is a decrease of F due to the increase in the intake momentum drag term $\dot{m}V$ in Equation (5.1). At low forward speeds, i.e. during the take-off ground run, this is the dominant effect but as V increases the second effect becomes important, the increase of ram pressure at entry to the fan or compressor. This occurs because the air does not enter the engine at the full flight speed; its velocity relative to the engine is reduced to a much lower value at entry to the fan or compressor and this reduction of velocity generates a ram pressure which increases with aircraft speed. The increase of ram pressure increases the effective overall pressure ratio of the engine and this gives an increase of thrust as explained by Cohen et al. (1987).

ESDU 76034 (1976) gives generalised data for the variation of thrust with forward speed in the range of speeds used for take-off. The data are consistent with the equation

$$F/F_0 = 1 - k_2 V + k_3 V^2, \qquad (5.8)$$

where F_0 is the static thrust (i.e. at zero speed) and k_2 and k_3 are positive constants. Torenbeek (1982) has given empirical expressions for k_2 and k_3 and has shown that they both increase with the by-pass ratio λ. The Equation (5.8) is of course consistent with the functional relation (5.5) and is also consistent with the physical explanation given earlier, since the increase of intake momentum drag may be expected to give a nearly linear decrease of F with increasing V, while the effect of increasing ram pressure gives an increase of F that is roughly proportional to V^2. As an example, curve (1) in Figure 5.7 shows the variation of F/F_0 with speed in the ISA at sea level for the Rolls–Royce RB211-535E4 turbofan at the take-off rating. For speeds up to about 40 m/s the curve is nearly linear but at higher speeds there is some curvature due to the effect of the ram pressure. For the range of speeds shown the curve is fitted almost exactly by Equation (5.8) with $k_2 = 2.52 \times 10^{-3}$ (m/s)$^{-1}$ and $k_3 = 4.34 \times 10^{-6}$ (m/s)$^{-2}$. For two other Rolls–Royce turbofans, the Tay 650 and the RB211-524G, the equation also gives a good fit with suitably chosen values of k_2 and k_3. The values of k_2 are found to be nearly the same as for the 535E4 but since k_3 represents only a small correction to the linear law its variation between different engines is relatively large. The by-pass ratio λ at take-off is 3.1 for the Tay 650 and 4.3 for the RB211-535E4 and 524G.

As the flight Mach number rises above about 0.4 the effect of increasing ram pressure at the fan or compressor entry becomes more important and the rate of decrease of thrust with increasing M

becomes smaller. As an example, Figure 5.8 shows the variation of F/F_0 with Mach number and height for the Rolls–Royce RB211-535E4 turbofan. Here F is the thrust at the cruise or climb rating at the specified height but the reference static thrust F_0 refers to the take-off rating at sea level. Since civil aircraft normally cruise at heights above 9 km and at Mach numbers between 0.75 and 0.85, Figure 5.8 shows that for this engine it is acceptable for approximate performance calculations to assume that in cruising conditions the maximum thrust is independent of speed at a given height. Data provided by Rolls–Royce plc show that this assumption is also valid for the Tay 650 and RB211-524G engines and the assumption is further supported by an empirical equation given by Mattingly, Heiser & Daley (1987) for turbofans with high by-pass ratio, showing that there is little variation of thrust as the Mach number rises from 0.75 to 0.85.

The assumption is also sometimes made, as in § 4.2, that for a given height the maximum thrust is independent of speed during the climb, but Figure 5.8 shows that for this typical civil turbofan engine this assumption can give only a rough approximation. Taking as an example an aircraft with $V_e^* = 100$ m/s, climbing at a height of 3 km with $\epsilon = 4$, the value of v shown in Figure 4.3 for maximum rate of

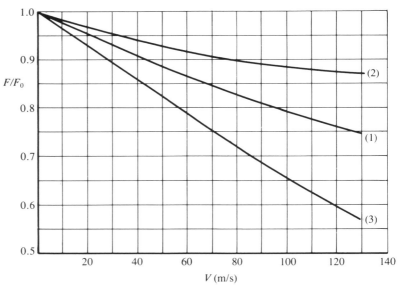

Figure 5.7. Variation of maximum take-off thrust with forward speed at sea level.
    (1)   Rolls–Royce RB211-535E4 turbofan.
    (2)   Military turbofan 'A', no reheat.
    (3)   Propfan 'C'.

Turbojet and turbofan engines

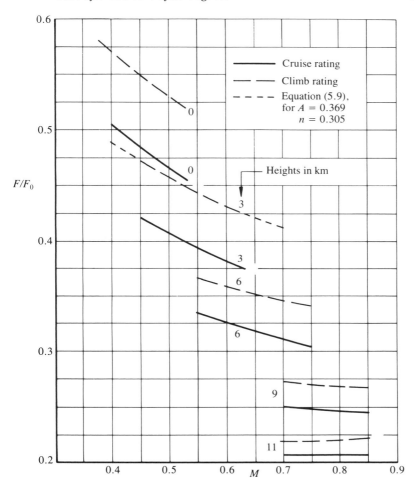

Figure 5.8. Variation of maximum thrust with height and Mach number. Rolls–Royce RB211-535E4 turbofan.

climb is about 1.7, the Mach number is then about 0.6 and Figure 5.8 shows that for this engine there would be a significant decrease of thrust with increasing speed. A much better approximation is obtained by assuming that

$$F/F_0 = AM^{-n}, \qquad (5.9)$$

where A and n are positive constants, and the curve plotted from this equation in Figure 5.8 shows that with $A = 0.369$ and $n = 0.305$ there is excellent agreement with the climb thrust curve over a limited range of speed at a height of 3 km. Equally good agreement can be obtained with the climb thrust curves for this engine at other heights, using $n = 0.326$ for $h = 0$ and $n = 0.223$ for $h = 6$ km, and also for the Rolls–Royce Tay 650 and RB211-524G. It may therefore be concluded that for civil turbofans the variation of climb thrust with speed at a given height may be represented with good accuracy over a limited range of speed by Equation (5.9), with suitably chosen constants.

The other important variable affecting the maximum thrust of a turbofan is the flight altitude. The functional relation (5.5) shows that if M and $N/\theta^{1/2}$ remain constant and if effects of varying Reynolds number within the engine can be neglected, the thrust varies in direct proportion to the air pressure as the height is varied. For heights between 11 and 20 km in the ISA the temperature is constant, so that if the engine speed is kept constant $F \propto \delta$, for a given flight Mach number. In the range of heights from 0 to 11 km in the ISA the reduction of temperature with increasing height gives an increase of $N/\theta^{1/2}$ if the engine speed is kept constant and this gives a large increase of F/δ. As an example, Figure 5.9 shows the variations of F/F_0 and $(F/F_0)\delta^{-1}$ at a constant Mach number of 0.7 for the Rolls–Royce RB211-535E4, where F is measured at the cruise rating and F_0 is the sea-level static thrust for take-off, as before. It can be seen that the thrust F decreases with increasing height but for

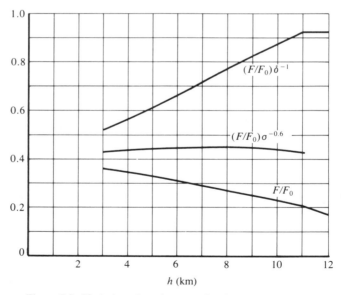

Figure 5.9. Variation of maximum cruise thrust with height at $M = 0.7$. Rolls–Royce RB211-535E4 turbofan.

Turbojet and turbofan engines

$h < 11$ km the decrease is much smaller than it would be if F/δ were constant.

The empirical equation given by Mattingly *et al.* (1987) that was mentioned earlier shows that for turbofans with high by-pass ratio $F/F_0 \propto \sigma^{0.6}$, for any constant Mach number up to 0.9, and the validity of this for the 535E4 engine at $M = 0.7$ is confirmed by the middle curve of the three given in Figure 5.9, showing that $(F/F_0)\sigma^{-0.6}$ is nearly constant. Further investigation has shown that this result is also true for other civil turbofans.

5.2.3 FUEL CONSUMPTION OF CIVIL TURBOFANS

The functional relation (5.7) shows that for fixed values of the engine speed and air temperature the sfc may vary with Mach number. It is found that the sfc does increase with Mach number, especially when the by-pass ratio λ is large, and this is illustrated in Figure 5.10 which refers to the Rolls–Royce RB211-535E4 at the cruise rating, for which $\lambda = 4.3$. In this figure the ordinate is the ratio of the specific fuel consumption c at the specified height and Mach number to the value c_{so} in static conditions (i.e. $V = 0$) at the take-off rating at sea level. Two forms of simple empirical law are often used to represent the

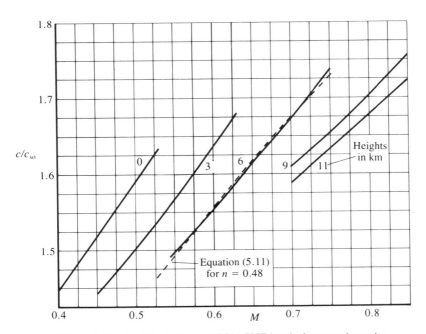

Figure 5.10. sfc of Rolls-Royce RB211-535E4 turbofan at cruise rating.

variation of sfc with Mach number. The first is a simple linear law

$$c = c_0(1 + k_4 M), \qquad (5.10)$$

where c_0 and k_4 are constants, and at constant height this usually agrees well with measured data over the range of M that is considered for cruising flight (say $0.7 < M < 0.85$). All the curves shown in Figure 5.10 are nearly straight and are therefore well represented by Equation (5.10), but it should be noted that the equation is valid over only a limited range of Mach number and the constant c_0 is not the value of c at zero speed.

An alternative law that is more convenient for analysis of range and fuel consumption is a power law based on the functional relation (5.7) viz

$$c = c_1 \theta^{1/2} M^n, \qquad (5.11)$$

where c_1 and n are constants. The broken line in Figure 5.10 shows that, with $n = 0.48$ and an appropriate choice of c_1, this equation gives good agreement with the measured data for this engine at a constant height of 6 km. With suitably chosen values of c_1 and n, equally good agreement can be obtained over a limited range of Mach number for this engine at other heights and for other civil turbofan engines.

The functional relation (5.7) shows that Equation (5.11) may be expected to give a good approximation over a limited range of M even if a change of height causes a variation of θ, provided that $N/\theta^{1/2}$ is kept constant. In other words, if the temperature drops due to an increase of height, the Equation (5.11) remains reasonably accurate if the engine speed N is correspondingly reduced. ESDU 73019 (1982) goes further than this and suggests that, even if $N/\theta^{1/2}$ is not kept constant, the equation gives a reasonable approximation to c within the limited range of N, θ and M likely to be used in cruising flight. Figure 5.11 has been drawn as a test of this more general validity of Equation (5.11), noting that $n = 0.48$ gives good agreement with measured data for the 535E4 engine at a height of 6 km. If the equation is generally valid as suggested in ESDU 73019 (1982), $\theta^{-1/2}(c/c_{so})M^{-n}$ should be constant and this quantity is plotted against Mach number in Figure 5.11 for several heights, taking n to be 0.48. The figure shows that for heights between 3 and 11 km and for the range of Mach numbers shown the variation of $\theta^{-1/2}(c/c_{so})M^{-n}$ from the mean value is no more than $\pm 1.6\%$ and even if the sea-level curve is included the variation is only $\pm 2.5\%$. Thus Figure 5.11 confirms that Equation (5.11) does give a reasonably good approximation to the variations of c due to changes of both Mach number and height.

ESDU 73019 (1982) also gives values of n for some current (1982) turbofan engines at their cruise ratings. The data apply to Mach

Table 5.1. *Mean values of n in Equation (5.11) for h = 9 km and 0.6 ≤ M ≤ 0.9, from ESDU 73019 (1982).*

λ	0	2	4	6	8	10
n	0.17	0.28	0.36	0.46	0.52	0.57

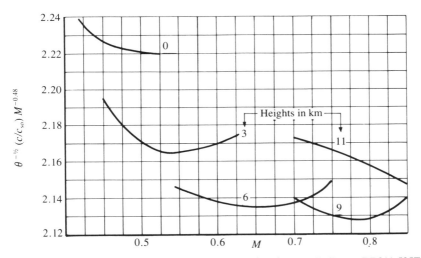

Figure 5.11. Test of validity of Equation (5.11) for Rolls-Royce RB211-535E4 turbofan at cruise rating.

numbers from 0.6 to 0.9 at a height of about 9 km. There is considerable scatter in the data but the mean values are as given in Table 5.1, showing an increase with the by-pass ratio λ. For the RB211-535E4 considered here the value of n giving the best agreement with the measured data at a height of 9 km is 0.45, although as noted earlier it is 0.48 for $h = 6$ km. This value of 0.45 is higher than that indicated in Table 5.1 for this engine with $\lambda = 4.3$, but 0.45 is within the scatter of the data given in ESDU 73019 (1982).

If the pilot chooses to reduce the thrust below its rated value by movement of the throttle lever, there is a consequent change of sfc which may sometimes be important. Examples showing typical behaviour are given in Figure 5.12, where the sfc ratio c/c_{CR} is plotted against the thrust ratio F/F_m for two Rolls–Royce civil turbofans. Here c_{CR} and F_m are respectively the sfc and the thrust at the cruise rated condition. For both the engines represented in the figure, the minimum sfc at the typical cruise condition of $M = 0.8$ and $h = 11$ km is obtained when the thrust is reduced to about 75–80% of the rated value, and this reduction of thrust gives about 2–3% reduction of sfc. At lower Mach numbers and heights there is no such reduction of sfc as the thrust is reduced and this is shown by the curves given in Figure 5.12 for $h = 6$ km and $M = 0.65$. In all cases there is a substantial increase of sfc as the thrust falls to a low value, showing that operation at a fairly high thrust can be beneficial in improving efficiency. For the upper pair of curves shown, there is a drop of about 5% in sfc when the thrust is increased by one third from 0.6 to 0.8 of the rated cruise thrust.

5.2.4 MILITARY TURBOFANS AND PROPULSION FOR SUPERSONIC CIVIL AIRCRAFT

The engines normally used for high performance military aircraft are turbofans having by-pass ratios between 0 and 1, with provision for reheat. As mentioned earlier, reheat is often used for take-off, for acceleration and for combat manoeuvres and is sometimes used for all supersonic flight. It is a valuable feature of these military turbofans because it does not increase the frontal area and adds little to the weight, yet it enables the thrust to be greatly increased. There is an associated increase of sfc, but this can be accepted if the reheat is used only for short periods. Reheat is also used in the Olympus 593 engines of Concorde, for take-off and for the accelerating climb through sonic speed up to $M = 1.7$.

The typical performance of a military turbofan will be illustrated by

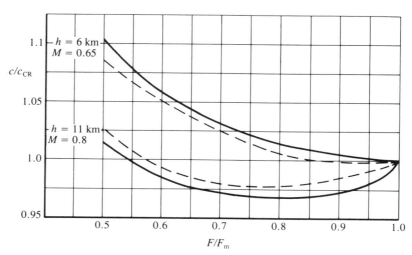

Figure 5.12. sfc of civil turbofans at reduced thrust. ——— Rolls–Royce Tay 650. – – – Rolls–Royce RB211-524G.

giving data for a turbofan 'A', which is representative of an engine suitable for use in a high performance military aircraft. Following the usual practice, the term 'dry' will be used to denote conditions when no reheat is used, because there is then no supply of fuel to the afterburner. The term 'maximum dry' denotes the condition of maximum thrust in the dry state and 'maximum reheat' refers to the maximum amount of reheat that can be used without exceeding the limiting HP turbine stator outlet temperature that is specified for the maximum dry condition.

For the range of speeds used for take-off, the curve (2) in Figure 5.7 shows the variation of thrust of the dry turbofan 'A' with forward speed. As for the civil turbofan represented by curve (1), the curve (2) is specified almost exactly by Equation (5.8), but the appropriate values of k_2 and k_3 are now 1.77×10^{-3} (m/s)$^{-1}$ and 6.1×10^{-6} (m/s)$^{-2}$ respectively. The constant k_2 is smaller for the military engine than for the civil one, because the by-pass ratio is lower and hence for a given thrust the mass flow rate $\dot{m}$ and the intake momentum drag are reduced. (The value of k_2 is even smaller when reheat is used.)

In Figure 5.13 the ratios F/F_0 and c/c_{so} for the turbofan 'A' in the maximum dry state are plotted against Mach number. Here F_0 and c_{so} both refer to the maximum dry state at sea level and zero forward speed. As the speed increases from a low value the form of the thrust variation changes from a decrease, as shown in Figure 5.7, to an increase as shown in Figure 5.13, because of the effect of increasing ram pressure at entry to the engine. For heights above 6 km the thrust increases with speed for all Mach numbers between about 0.4 and 1.4 or more, in contrast to the civil turbofan behaviour shown in Figure 5.8. At high supersonic speeds the thrust drops sharply with increasing speed and would fall to zero at some Mach number between 2 and 3. Equation (5.1) shows the reason for this; the thrust falls to zero when the flight speed V reaches the mean jet velocity V_J and the latter velocity, being governed more by the combustion conditions than by the speed of the intake air, is relatively unaffected by V.

For civil turbofans it has been shown that the variation of thrust with speed can be represented well by the power law of Equation (5.9), but for the military turbofan the range of Mach number to be considered is much greater and a simple power law representation is not useful.

As explained in § 5.2.1, for constant values of the engine speed and flight Mach number the variation of thrust with height should be given by $F \propto \delta$ for heights between 11 and 20 km, where the air temperature is constant in the ISA. This is equivalent to $F \propto \sigma$ in this height range and the relation is confirmed approximately for the whole range of Mach number by comparison of the data for 11 and 15 km in Figure

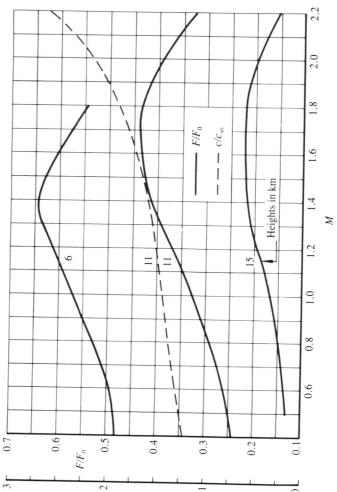

Figure 5.13. Thrust and sfc of military turbofan 'A' in maximum dry condition.

5.13. Moreover the data for lower heights show that, for this particular engine in the subsonic speed range, F is roughly proportional to σ for all heights above about 6 km. In the supersonic speed range, if the variation of thrust with height is expressed as $F \propto \sigma^n$ for $h < 11$ km, the values of n required to fit the data are less than 1 and decrease as the Mach number rises.

In Figure 5.13 the ratio c/c_{so} is shown only for the one height of 11 km because at all Mach numbers the variation of c with height is quite small, no more than about ±3% for $M < 1$ and ±5% at the highest Mach numbers. For this engine the variation of c with M is almost linear for $M \leq 1.4$, but at higher Mach numbers there is a more rapid rise of c. Equation (5.11) could be used to represent the variation of c with M over any specified *small* range of M, but the constants in the equation would depend on the selected range.

Figure 5.14 refers to the turbofan 'A' in the maximum reheat condition and shows the same ratios as Figure 5.13. (The reference quantities F_0 and c_{so} are the same as in Figure 5.13 and refer to the maximum *dry* state.) Both the thrust and the sfc are greatly increased by the reheat and there is now no decrease of thrust at the higher Mach numbers, because the reheat increases the mean jet velocity V_J. As for the dry engine the variation of thrust with speed cannot be represented usefully by the power law of Equation (5.9). In the subsonic speed range the variation of thrust with height is given approximately by $F \propto \sigma^n$ for heights between 6 and 11 km, but the required value of n is about 1.3 and not 1.0 as it was for the dry engine.

For the turbofan 'A' with maximum reheat the variation of c with height is quite small, as it was for the dry engine, and in Figure 5.14 the ratio c/c_{so} is shown for only one height. The form of the curve giving c/c_{so} is complex and it is notable that the large increase shown at high Mach numbers for the dry engine is no longer present.

Figure 5.15 shows in another form the effects of reheat on thrust and sfc. The suffixes DRY and RH refer to the maximum dry and maximum reheat states defined earlier and the ratios shown therefore give the factor by which the maximum dry values of F and c are multiplied when maximum reheat is used. At Mach numbers above about 2 the thrust factor becomes very large because the dry thrust is decreasing towards zero as the flight speed V increases towards the jet velocity V_J. In this range of Mach number the sfc factor becomes relatively small, showing that the greatly increased thrust is obtained with only a modest increase of sfc, although the fuel flow rate Q is large because $Q = cF$.

Figure 5.16 shows how the sfc of the dry engine changes as the thrust is reduced by the pilot's use of the throttle control. The

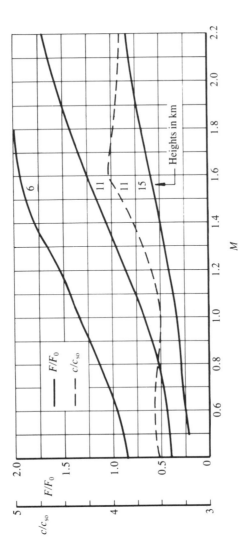

Figure 5.14. Thrust and sfc of military turbofan 'A' in maximum reheat condition.

Turbojet and turbofan engines

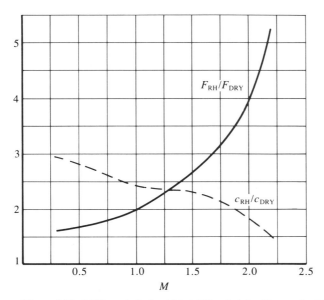

Figure 5.15. Military turbofan 'A' at 11 km height. Factors for increase of thrust and sfc due to maximum reheat.

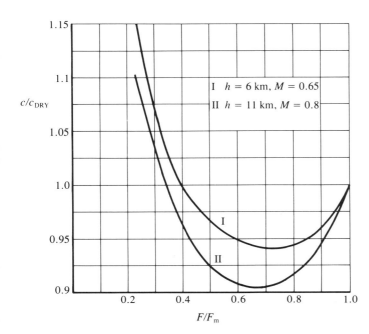

Figure 5.16. sfc of military turbofan 'A' at reduced thrust.

maximum dry thrust is denoted here by F_m and the sfc at this thrust is c_{DRY}. The curves are directly comparable with those in Figure 5.12 for civil turbofans and show that for the military engine the minimum sfc is obtained when F is about 70% of the maximum value. As for the civil turbofans the reduction of sfc becomes greater as the height and Mach number are increased.

For a high performance military aircraft one of the main requirements is usually low fuel consumption in a cruise at a subsonic speed and this has an important influence on the selection of a suitable engine. In future designs of supersonic civil aircraft this may also be true, because of the limitations imposed by the supersonic boom, but for Concorde the primary requirement was low sfc in cruising flight at $M = 2$ and fuel economy at subsonic speeds was considered to be of secondary importance. This led to the choice of the Rolls–Royce Olympus 593 engine, a pure turbojet with no by-pass flow and with provision for reheat. Figure 5.17 shows some characteristics of this engine, presented in accordance with the conclusions noted in § 5.2.1 that, for heights above 11 km and for fixed values of the engine speed and flight Mach number, F/δ and c are independent of height. The curves refer to the dry engine at maximum cruise thrust and show how F/δ and c vary with Mach number at any height above 11 km. For this engine F/δ increases almost linearly with Mach number for $1 < M < 2$ and in this range of speed there is little change of c. At Mach numbers above 2 the ratio F/δ would fall for the reason explained earlier in discussion of Figure 5.13 and some increase of c would also be expected, as shown for high Mach numbers in Figure 5.13.

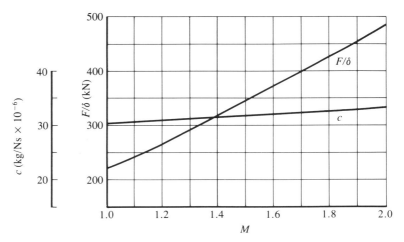

Figure 5.17. Thrust and sfc of Olympus 593 turbojet at cruise rating, no reheat, $h > 11$ km.

The maximum reheat used in the Olympus 593 engine increases both the thrust and the sfc at supersonic speed by about 25–30%, much less than the values shown for the military turbofan 'A' in Figure 5.15. The requirements for the two engines are of course quite different; reheat is needed in the Olympus only to provide sufficient thrust for take-off and for acceleration through sonic speed, whereas it is used on a military aircraft to give very high speed and rate of climb, even in a turn.

5.3 Propellers

Propellers are used on nearly all light aircraft in private use and also on a considerable range of passenger and freight transport aircraft where high flight speed is not required. In these cases the high propulsive efficiency of a propeller in comparison with a turbofan can give a considerable benefit. For the low powers required for light aircraft small piston engines are used to drive the propellers, but for powers above about 300 kW gas turbines are always used because they have greater power/weight ratios than piston engines. The essential features of propellers are independent of the power source that is used to drive them and these will be considered in this section. Gas turbines driving propellers, known as *turboprops,* will be considered in § 5.4 and some other forms of power plant, using open rotors and maintaining high efficiency up to high subsonic flight speeds, will be discussed in § 5.5. Finally, § 5.6 will give a brief survey of the essential characteristics of piston engines used to drive propellers.

The thrust F_P developed by a propeller and the shaft power P_s transmitted to it are usually expressed in dimensionless form as

$$\textit{Thrust coefficient} = C_T = F_P/(\rho N^2 d^4) \tag{5.12}$$

and

$$\textit{Power coefficient} = C_P = P_s/(\rho N^3 d^5), \tag{5.13}$$

where d is the diameter of the propeller and N is its rotational speed (rev/s).

The *efficiency* of a propeller η_{PR} is defined as the ratio of the useful thrust power to the shaft power so that

$$\eta_{PR} = F_P V/P_s = JC_T/C_P, \tag{5.14}$$

where $J = V/(Nd)$ and is known as the *advance ratio*. The value of J is low when the air stream passes through the propeller disc relatively slowly while the rotational speed N is relatively large, e.g. at the start of the take-off ground run when the aircraft velocity V is small and the thrust is high. In contrast, J is high when the air flows rapidly through the disc and N is relatively low, as in a high-speed dive with low thrust.

Dimensional analysis shows that for propellers of fixed geometrical shape (and fixed blade angle) C_T, C_P and η_{PR} are functions of J and of appropriately defined Reynolds and Mach numbers. Over the operating range of a given propeller the influence of varying Reynolds number is usually negligible, but effects of varying Mach number are often important. It is usual to define a Mach number M_T corresponding to the resultant velocity of the blade tip relative to the air, neglecting the components of induced velocity, so that

$$M_T = [V^2 + (\pi N d)^2]^{1/2}/a = M[1 + (\pi/J)^2]^{1/2} \tag{5.15}$$

where $M = V/a$, the flight Mach number. The effects of varying Mach number are usually small for values of M_T les than about 0.9 but as M_T rises above that value an increasing portion of each blade from the tip inward becomes subject to the rapid increase of drag associated with the formation of shock waves. This usually leads to a loss of efficiency but the loss is quite small on some modern propellers having very thin blade sections near the tips. High efficiency can also be maintained at large values of M_T by using blades which are swept back near the tips and this will be considered in § 5.5.

With reference to Figure 5.1 and Equation (5.1), it can be shown that if the increase of velocity $(V_J - V)$ is uniform over the propeller disc the mass flow rate through an isolated propeller of disc area A is

$$\dot{m} = \tfrac{1}{2}\rho A(V + V_J),$$

so that

$$2F_P/(\rho A V^2) = (V_J/V)^2 - 1. \tag{5.16}$$

It follows that for a given flight speed and height the ratio of exit to inlet velocities can be reduced by reducing the *disc loading* F_P/A and this increases the propulsive efficiency η_p as given by Equation (5.2). Thus an increase of propeller diameter always gives an increase of η_p for a given thrust and it is found that the overall propeller efficiency η_{PR} also increases, although η_{PR} is always less than η_p because of blade drag and the loss of energy associated with rotation of the slipstream (unless contra-rotating propellers are used).

In choosing a propeller design the advantage of large diameter in increasing efficiency needs to be balanced against the disadvantages of reduced tip clearance from the ground or fuselage, stiffness problems of long slender blades and the need for a lower rotational speed unless an increase of tip speed can be accepted.

Equation (5.14) shows that the propeller efficiency η_{PR} is zero when J or V is zero, because the thrust power $F_P V$ is then zero. As J increases from a low value the efficiency for a fixed blade angle increases up to a maximum of about 0.88 or 0.90, for a propeller with low disc loading, and then decreases as the mean angle of incidence and mean lift coefficient of the blades decrease, so that the blades carry progressively less load until eventually both the thrust and the efficiency fall to zero. This condition of zero thrust might occur, for example, in a dive if the propeller shaft speed did not increase substantially, while the flight speed did increase, thus raising J. The maximum efficiency decreases with increase of disc loading, to some extent for the reason already given but mainly because of the increased loss due to rotation of the slipstream. By the use of contra-rotating propellers this rotational loss can be eliminated and maximum efficiencies approaching 0.9 can be obtained even with high disc loading.

On all propeller driven aircraft except those of very low power the blade angle is controlled in flight to keep the rotational speed N at a specified constant value. The efficiency then remains high over a range of J that covers the flight speeds used for climbing and cruising and this is illustrated in Figure 5.18, which is derived from data given by Hamilton Standard for a 4-blade propeller, reproduced by Torenbeek (1982) and by Lan & Roskam (1981). The broken line in Figure 5.18 shows how η_{PR} would vary with J if the blade angle (measured at $\frac{3}{4}$ of the tip radius) were fixed at 40°, while the full line shows η_{PR} for a constant power coefficient of 0.3. With the blade angle controlled in flight to keep N constant, a constant value of C_P represents constant shaft power. The curves show that the range of J for which η_{PR} is high is greatly extended by varying the blade angle, so that at $C_P = 0.3$ the efficiency exceeds 0.83 for a range of J from 1.3 to 2.7, instead of only 1.6–2.1 for the constant blade angle shown. At constant N the range of J at constant C_P represents a 2:1 variation of forward speed.

The discussion so far has referred to an isolated propeller, but when the propeller is mounted on an aircraft there are important interference effects to be considered. If the propeller is a 'tractor', i.e. it is mounted on the aircraft ahead of a fuselage or engine nacelle, there are two influences on effective thrust. First, the presence of a body close behind the propeller reduces the axial component of relative air velocity, so that for given Nd the effective value of J is less than $V/(Nd)$, where V is the flight speed, and this leads to an increase of thrust. The second influence is on drag rather than thrust; the increase of relative air velocity behind the propeller leads to an increase in the drag of the fuselage or engine nacelle and any part of the wing that is within the slipstream. The overall effect of these two influences is usually a loss of effective thrust and this reduces the propeller efficiency by about 0.02 or 0.03 in comparison with the free-air value. Methods of estimating the effects of the propeller installation have been given by Bass (1982) and by Torenbeek (1982). When the propeller is mounted as a 'pusher', i.e. behind an engine nacelle, the effects of the installation on the propeller efficiency are more difficult to estimate but Bass (1982) notes that pushers are usually more efficient than tractors.

Provided the tip Mach number M_T is not too high the propeller efficiency remains nearly constant in the range of flight speeds used for climbing and cruising, so that the thrust power $F_P V$ is nearly proportional to the shaft power output of the engine. As already noted, the propeller efficiency falls to zero at zero flight speed and in estimating take-off performance allowance must be made for the reduction of propeller efficiency at low forward speeds. Empirical curves for estimating propeller thrust at low forward speeds are given in ESDU 83001 (1983) and these show that in a typical case the propeller efficiency is reduced from the cruising value of 0.85 to about 0.6 as the forward speed falls to about 35 m/s.

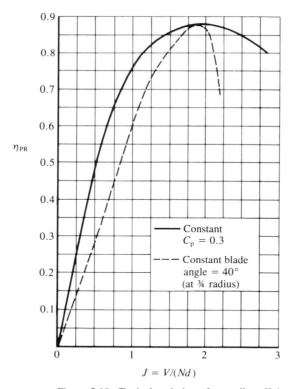

Figure 5.18. Typical variation of propeller efficiency with advance ratio. Hamilton Standard data reproduced by Torenbeek (1982) and by Lan & Roskam (1981).

For a typical turboprop the large diameter of the propeller makes it equivalent to a turbofan engine with a by-pass ratio in the region of 50. In cruising flight the ratio V_J/V in Equation (5.2) may be less than 1.02, giving a propulsive efficiency that is greater than 0.99. The overall propeller efficiency η_{PR} may be as high as 0.88 if the disc loading is low, but this may be reduced to about 0.85 by the interference effects discussed earlier. This high efficiency can give a fuel consumption that is some 25% lower than that of a modern turbofan engine (with a by-pass ratio in the range 3–6) but it is necessary to ensure that the blade tip Mach number M_T is not too high and attention must now be given to the acceptable upper limit of M_T.

For older designs of propeller the increase of blade drag in the tip region at high values of M_T led to severe loss of efficiency as M_T increased above about 0.9. For modern propellers the situation is different; the ratio of blade thickness to chord is as low as 0.02 in the tip region and the disc loading is usually greater than it was for the older propellers. As a result, the total loss of efficiency due to blade drag is only about 4% and even a large increase of blade drag due to formation of shock waves near the blade tips reduces the efficiency by only about 2%. For these modern highly loaded propellers the main contribution to loss of efficiency comes from the lost energy in the rotating slipstream and this can be eliminated by the use of a contra-rotating pair.

The same shock waves that cause an increase of blade drag at high values of M_T also produce a very loud noise and it is for reasons of noise limitation, rather than efficiency, that with modern propellers M_T is usually restricted to an upper limit of about 0.85, or possibly 0.9. It is the noise in the cabin of the aircraft, not on the ground, that causes the main problem because it is only in the later stages of the climb and in the cruise at high altitude that M_T may become large. Since most of the noise is radiated in the plane of the propeller disc the problem becomes less severe when rear-mounted pusher propellers are used, because the greater part of the cabin is then ahead of the region of high noise.

Assuming that M_T is to be restricted to an upper limit of 0.85 or 0.9, there is a corresponding upper limit of flight Mach number which will be denoted by the symbol M_P. Since the rotational velocity component πNd in Equation (5.15) is usually at least as great as the cruising flight speed the value of M_P is usually not much greater than 0.6. This imposes a serious handicap on propeller-driven aircraft, a price to be paid for the low fuel consumption. Propfans and other advanced forms of open-rotor power plant use very thin blades which are curved, so that the tip regions are swept back, and this alleviates the noise problem as well as reducing the rate of rise of blade drag with increasing M_T. These power plants will be discussed in § 5.5, where it will be shown that they allow cruising Mach numbers as high as 0.8 with very good fuel economy, although there may still be a problem with noise.

5.4 Turboprops

In the design of a turboprop, a gas turbine driving a propeller, the modern practice is to design the gas turbine power plant and the propeller together as a single unit. The manufacturer then offers to the aircraft designer a complete unit with characteristics that are specified in terms of thrust rather than power and with the sfc expressed as $c = Q/F$ rather than $c' = Q/P_e$. Nevertheless much of the discussion of turboprops to be given here will be based either on the shaft power P_s or on the equivalent shaft power P_e, to be defined later. The sfc will be defined either as $c_s' = Q/P_s$ or as $c' = Q/P_e$, because these definitions allow some useful approximations to be made, e.g. that c' or c_s' at the cruise rating is independent of flight speed for a given height.

For a conventional turboprop the upper limit of cruising Mach number M_P has already been mentioned. A further adverse feature of the turboprop is its relatively high weight per unit of cruise thrust, caused by the weight of the propeller and reduction gear, and in considering the use of a turboprop rather than a turbofan the disadvantages of low cruising speed and low thrust/weight ratio must be balanced against the advantage of good fuel economy.

In a turboprop power plant the gas stream leaving the turbine gives some direct thrust F_J, but for modern turboprops this is usually no more than 5% of the thrust F_P given by the propeller. It is convenient to make allowance for the jet thrust by defining an *equivalent shaft power* P_e as the shaft power that would be needed to give a propeller thrust equal to the true total thrust. Thus

$$\eta_{PR} P_e = V(F_P + F_J) = \eta_{PR} P_s + F_J V \qquad (5.17)$$

or

$$P_e = P_s + F_J V / \eta_{PR}, \qquad (5.18)$$

where P_s is the true shaft power transmitted to the propeller. The equations become indeterminate in the static condition when V and η_{PR} are zero, but Equation (5.17) can be rearranged to show that

$$F_P / P_s = \eta_{PR} / V$$

and although the right-hand side of this equation is indeterminate when $V = 0$ the left-hand side can at least be assigned a typical value. Indeed it is usual to establish the Equation (5.18) for the static condition by using a typical value of F_P/P_s for a propeller. The value

Turboprops

given by Saravanamuttoo (1987) which has often been used is 2.5 pounds force per horsepower or 14.9 N/kW. Thus in the static condition, with powers measured in kilowatts and F_J in newtons,

$$P_e = P_s + F_J/14.9. \tag{5.19}$$

On a turboprop with the rotational speed controlled automatically at a preset value the shaft power can be varied over a wide range by adjustment of the propeller blade pitch. The pilot normally has two controls for the power plant, a speed control for setting the rotational speed and a power control for varying the shaft power. When the pilot demands an increase of power the propeller adjusts to a coarser pitch and the fuel flow increases to maintain the preset rotational speed. The upper limit of power is expressed in terms of ratings for take-off, climb and cruise and the rated power for each of these is the maximum available. This rated power is limited by various factors in the design of the engine, of which the most important is usually the maximum permissible gas temperature at inlet to the turbine, and the controller for fuel flow ensures automatically that this limit is not exceeded.

5.4.1 MAXIMUM SHAFT POWER

The maximum available shaft power of a turboprop engine increases with flight speed because of the increase of ram pressure in the engine intake. As mentioned in § 5.2.2 the increase of ram pressure tends to increase the thrust of a turbojet or turbofan engine and correspondingly for a turboprop it increases the shaft power. The increase of shaft power with speed is shown, for example, in Figure 5.19, which refers to a typical modern small turboprop 'B' at the climb rating. The quantity plotted as ordinate is the ratio of the shaft power P_s at either of the two specified heights to the shaft power P_{so} at the take-off rating at sea level. This engine is 'flat rated' at low altitudes, i.e. it is controlled in such a way that the shaft power at either the climb or the cruise rating remains constant over a wide range of speed and for all heights up to at least 3 km. (The power P_{so} at the take-off rating at sea level also remains constant over a wide range of speed.)

The climb rating has been selected for the example of Figure 5.19 because the variation of power with flight speed is of interest mainly in connection with climbing performance, as discussed in Chapter 4. The points shown in the figure are derived directly from the engine data and the lines represent the equation

$$P_s/P_{so} = AM^n, \tag{5.20}$$

which is of course closely analogous to Equation (5.9), but the constants A and n now have different meanings. With appropriately chosen values of A and n the equation gives excellent agreement with the data for this engine and the required value of n is close to 0.5 for both the heights shown. For other turboprop engines the Equation (5.20) is also likely to give good agreement with measured data, but the required values of n may be different, although they will always be between 0 and 1.

The propeller thrust power is

$$F_\text{P} V = \eta_\text{PR} P_\text{s} \tag{5.21}$$

and if η_PR is constant and $P_\text{s} \propto M^n$ the propeller thrust F_P at any given height is proportional to $M^{(n-1)}$, but a moderate variation of η_PR over a relatively small range of speed may alter this conclusion significantly. For example, taking the result shown in Figure 5.19 for a height of 9 km, $F_\text{P} \propto M^{-0.48}$ if η_PR is constant, but if η_PR decreases from 0.86 at $M = 0.55$ to 0.84 at $M = 0.65$ the thrust law in this speed range becomes $F_\text{P} \propto M^{-0.62}$.

As the height increases, the reduction of air density leads to a reduction of the mass flow rate of air through the engine and this causes a substantial reduction of maximum available shaft power,

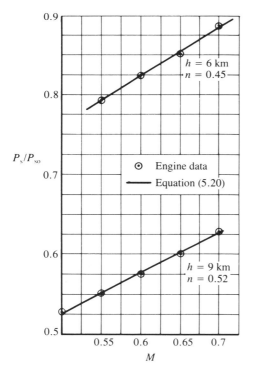

Figure 5.19. Shaft power of small turboprop 'B' at climb rating.

Turboprops

unless the engine is 'flat rated' in the range of height considered. The change of shaft power with height is illustrated for the small turboprop 'B' considered here by the difference between the two lines in Figure 5.19. In this case, for any fixed Mach number within the range shown, the ratio of the shaft powers for heights of 9 and 6 km is close to the ratio of air densities, viz 0.707, and this is also true if the powers are considered at the cruise rating instead of the climb. Thus for this particular engine the maximum available shaft power is directly proportional to the air density, but this simple relation is not generally applicable.

More generally, it may be assumed that the variation of maximum power with height is given by a law of the form $P_s \propto \sigma^n$, and it should be noted that the value of n will be smaller when the EAS is kept constant as the height varies than when M is kept constant. This is because an increase of height at constant EAS gives an increase of M, so that the power at the increased height is greater than it would be if M were kept constant.

So far, the discussion of turboprop power has been based on the shaft power P_s and not on the equivalent shaft power P_e as defined by Equation (5.17), which is more useful in performance calculations because it includes an allowance for the direct jet thrust. If it can be assumed that the ratio P_e/P_s is constant and if the shaft power is expressed as a multiple of a datum value, as in Figure 5.19, all the conclusions about the variation of P_s with speed and height apply equally well to P_e. The importance of the direct jet thrust F_J and its possible effect in causing a variation of P_e/P_s will now be considered.

Equation (5.18) shows that any variation of F_J with speed and height will lead to a change in the ratio P_e/P_s unless $F_J V/\eta_{PR} \propto P_s$. With increasing height at constant true air speed there is a reduction of F_J because of the reduced mass flow rate and, because P_s also decreases, it may be acceptable to assume that $F_J \propto P_s$ and that η_{PR} remains constant, so that there is no change of P_e/P_s. With increasing flight speed at a constant height there is a decrease of F_J, for the relevant range of Mach number up to about 0.6, because of the increase of intake momentum drag, and it may sometimes be reasonable to assume that $F_J V \propto P_s$, with constant η_{PR}, although this assumption is not likely to be accurate. Nevertheless, since jet thrust F_J is usually small in comparison with F_P, it may often be acceptable to assume that variations of F_J with both height and speed have negligible effects on the ratio P_e/P_s. Some support for the assumption is provided by data given by Hill & Peterson (1965) for the Rolls–Royce Tyne, showing that even for this early engine with F_J/F_P as high as 0.12 in the cruise, the ratio P_e/P_s does not vary by more than ±3% for speeds from 50 to 200 m/s and for heights from sea level up to 9 km. For a modern turboprop F_J/F_P would be no more than 0.05 in the cruise and the variation of P_e/P_s with speed and height would be smaller.

5.4.2 FUEL CONSUMPTION

As mentioned earlier the specific fuel consumption of a turboprop is often defined as

$$c' = Q/P_e, \qquad (5.22)$$

because this is more nearly independent of speed than the quantity $c = Q/F$ used for turbojet and turbofan engines. At the rated power for cruise or climb, an increase in the forward speed V at constant height gives an increase of P_s and P_e as discussed earlier, but the increase in the fuel flow rate Q is relatively small. Hence c' decreases with increasing speed, but the decrease is fairly small and over the range of speeds considered for cruising it may sometimes be acceptable to assume that c' is independent of speed, for a given height. When a more accurate representation is required the result given in Appendix A of ESDU 75018 (1975) may be used. This is based on a survey of current (1975) turboprop aircraft in cruising flight and shows that the variation with Mach number of the specific fuel consumption $c = Q/F$ (not $c' = Q/P_e$) may be represented with fair accuracy by Equation (5.11), with $n = 0.9$. Equations (5.17) and (5.22) give

$$c = Vc'/\eta_{PR},$$

so that

$$c'/\eta_{PR} = c/V \qquad (5.23)$$

and Equations (5.11) and (1.9) lead to

$$c'/\eta_{PR} = (c_1/a_0)M^{(n-1)}. \qquad (5.24)$$

If η_{PR} is assumed to be constant this leads to an equation for c' which may be written

$$c' = AM^{-m}, \qquad (5.25)$$

where A is a constant, $m = 1 - n$ and ESDU 75018 (1975) indicates that m should be about 0.1.

Figure 5.20 refers to the small turboprop 'B' considered earlier and shows the variation with M of $c_s' = Q/P_s$ (not $c' = Q/P_e$). The quantity plotted as ordinate is the ratio c_s'/c_{sTO}', where c_{sTO}' is the value of c_s' at the take-off rating in static conditions at sea level. Since $c_s' = c'P_e/P_s$, the quantity plotted is directly proportional to c' if it can be assumed that P_e/P_s is constant.

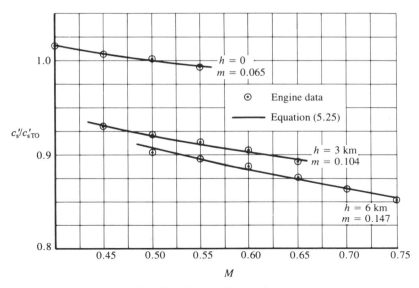

Figure 5.20. sfc of small turboprop 'B' at cruise rating.

The lines shown in Figure 5.20 represent the Equation (5.25), with values of A and m chosen to fit the data for this engine (with P_e/P_s assumed to be constant). The required value of m is 0.165 for $h = 9$ km and as shown on the curves for lower heights. Thus there is a progressive increase of m with height and some of the values differ significantly from the value of 0.1 suggested by ESDU 75018 (1975), although they do indicate that if a constant value of m is to be used for heights up to 6 km, 0.1 is a reasonable choice. Moreover, it should be noted that since m is small its exact value is not of great importance, e.g. with constant η_{PR} the reduction of c' as M increases from 0.5 to 0.6 is only 1.3% for $m = 0.07$ and 2.5% for $m = 0.14$.

At the rated cruise power the variation of c' or c_s' with height is usually small. For example, the data given by Roskam (1986) for the Pratt & Whitney PT6A-41 turboprop at the cruise rating show no appreciable change of c_s' as the height is increased from 6 to 9 km, at a constant true air speed of 150 m/s. Figure 5.21 shows a similar result for the turboprop 'B'; the variation of c_s' between 3 and 9 km is only about $\pm 1\%$, but there is a significant increase in c_s' as the height decreases to sea level.

When the pilot uses the throttle control to reduce the shaft power below the rated value, the operating condition of the engine moves away from the design point and there is an increase of c' and c_s' for any given flight speed and height. As an example, Figure 5.22 refers to the small turboprop 'B' and shows the variation with shaft power of the ratio c_s'/c_{sCR}', where c_{sCR}' is the value of c_s' at the rated cruise power P_{sm}. This shows, for example, that at a height of 3 km, with $M = 0.54$, a 30% reduction of power from the rated value increases c_s' by about 9%, and with constant P_e/P_s the percentage increase of c' would be the same.

If the power is kept constant, or nearly so, as height is reduced, the ratio P_s/P_{sm} decreases because the maximum power rises as shown (for the climb rating) in Figure 5.19. This causes a rise of c_s' as shown in Figure 5.22, an effect that is additional to the change of c_s' at the rated cruise power discussed earlier.

5.5 Propfans and other open-rotor power plants

As explained in § 5.2 a turbofan engine has a lower sfc than a comparable turbojet, because the by-pass flow from the fan reduces the mean jet velocity and hence increases the propulsive efficiency as given by Equation (5.2). The beneficial effect of the by-pass flow on the sfc increases with the by-pass ratio λ, but for very large values of λ (above about 8 or 10) the overall diameter of the power plant becomes large and the benefit of low sfc is lost because of the high drag and weight of the engine nacelle. In order to achieve a high by-pass ratio without the need for a large engine nacelle, various forms of open-rotor power plant have been developed, having a fan of one or two stages that is not enclosed within the nacelle. The conventional form of turboprop is of course one form of open-rotor power plant,

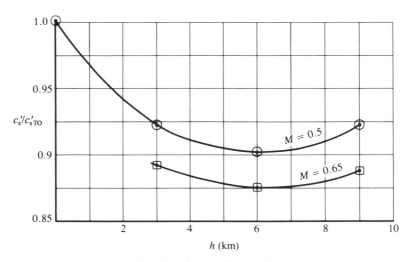

Figure 5.21. sfc of small turboprop 'B' at cruise rating.

Propfans and other open-rotor power plants

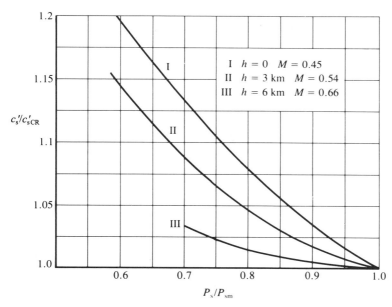

Figure 5.22. sfc of small turboprop 'B' at reduced power.

but as already explained the flight Mach number is limited in this case to a value in the region of 0.6, because at higher Mach numbers the shock waves that are formed near the blade tips lead to a high noise level and some loss of efficiency.

Open-rotor power plants of a new generation have been under development since the late 1970s by NASA and by most of the world's leading propeller and aero-engine manufacturers. These are generally known as propfans, but the General Electric Company* identifies its particular development as the UDF®, or unducted fan, engine†. In all these power plants the blades of the open rotors are very thin and highly cambered, with substantial sweepback at the outer radii, so that the formation of strong shock waves on the blades is delayed until high values of the blade tip Mach number M_T are reached. Thus high efficiency and acceptable levels of noise can be maintained up to flight Mach numbers as high as 0.8.

Figure 5.23 shows a NASA experimental open rotor mounted on the wing of a testbed aircraft. For the purposes of the experiments the rotor was mounted ahead of the gas turbine, like a conventional turboprop, but for practical use a rear mounted position is preferred,

---

* General Electric Company of the United States is not connected with the British company of a similar name.
† UDF® is a registered trademark of General Electric Company, USA.

110 *Power plants*

like a pusher propeller, in order to reduce the noise level in the aircraft cabin. In many designs the open rotor is driven from the gas turbine through a reduction gear, although this poses a considerable challenge to the gearbox designer, because the power is usually large and with a rear-mounted rotor the blades are close to the hot jet efflux. Moreover, for high efficiency the open rotor is best designed with two contra-rotating stages and this introduces a further complication in the design of the gearbox and rotor hub, although an alternative design has been proposed to avoid this complication, using a single rotor with fixed vanes on the downstream side to remove swirl.

Some of these difficulties are avoided in the UDF® engine de-

Figure 5.23. NASA advanced propeller installed on the wing of a Gulfstream II testbed aircraft.

Figure 5.24. An experimental UDF® engine installed on a testbed aircraft.

veloped by the General Electric Company and described by Gordon (1988). This is shown in an experimental form in Figure 5.24, with a pair of contra-rotating rotors mounted near the rear end of the power plant. The need for a reduction gear is avoided by driving the rotors directly from a pair of multi-stage contra-rotating turbines.

High efficiencies and acceptable noise levels have been reported at flight Mach numbers in the region of 0.8, both for gear driven propfans and for the UDF® engine. Whitlow & Sievers (1988) report an efficiency of 0.79 measured in a wind tunnel at $M = 0.8$ for a propfan with a single rotor and state that with a contra-rotating pair the efficiency can be improved by about 6%. Gordon (1988) states that the sfc of the UDF® engine is about 25–27% lower than that of the best turbofan engines going into service in 1988. He also quotes test results showing that there is no appreciable loss of efficiency as the flight Mach number increases up to 0.8 and only a small loss even at $M = 0.9$. The fuel savings that have been quoted for gear driven propfans are at least as good as those given here for the UDF® engine.

In considering the performance of a propfan or UDF® engine there are two possible approaches. The power plant with its open rotor may be regarded either as a turboprop, following the approach used in

112   *Power plants*

§ 5.4, or as a turbofan with a very high by-pass ratio (30–50). Much of the following discussion will be based on the latter approach, but it will be shown that some of the characteristics of a propfan are close to those of a turboprop.

5.5.1 MAXIMUM THRUST

In the earlier discussion of turboprops it was shown by consideration of Equation (5.21) that the propeller thrust decreases substantially with increasing flight speed. The behaviour of a propfan is similar and again there is a large decrease of maximum thrust as V increases.

In the range of speeds used for take-off it was shown earlier that the decrease of maximum thrust of a civil or military *turbofan* with increasing speed could be represented by Equation (5.8), with appropriate choice of the constants k_2 and k_3. The curve (3) in Figure 5.7 shows the variation of F/F_0 with speed in the ISA at sea level for a typical modern *propfan* 'C'. Equation (5.8) fits this curve well up to 130 m/s with $k_2 = 3.6 \times 10^{-3}$ (m/s)$^{-1}$ and $k_3 = 1.5 \times 10^{-6}$ (m/s)$^{-2}$, but a simple linear law with $k_2 = 3.5 \times 10^{-3}$ (m/s)$^{-1}$ and $k_3 = 0$ gives good agreement with the curve up to 100 m/s.

At higher flight speeds there is still a decrease of maximum thrust with increasing flight speed and this is shown, for example, in Figure 5.25, which refers again to the propfan 'C' and shows the variation of F/F_0 with Mach number and height, where F_0 is the static thrust at the take-off rating at sea level, as in the comparable Figure 5.8. This shows that for a propfan, unlike a turbofan, an assumption that F is independent of speed is not a good approximation, even in the typical cruise conditions of $M \simeq 0.8$ and $h \geq 9$ km. Either for cruise or for climb conditions the variation of F with speed for a propfan can be represented well by Equation (5.9), and it is found that for the propfan 'C' the best value of n in that equation is close to 0.7 for the cruise rating at all heights up to 9 km, and between 0.62 and 0.65 for the climb rating over the same range of heights. These values of n are substantially greater than those given earlier for civil turbofans and consideration of Figure 5.19, Equation (5.21) and the earlier discussion of that equation shows that the propfan 'C' behaves rather like a turboprop for which there is some decrease of η_{PR} with increasing speed.

With increasing altitude the maximum thrust of a propfan decreases, like that of all other forms of gas turbine power plant, and this is shown in the lower curve of Figure 5.26, for the propfan 'C' at a constant Mach number of 0.7. The upper curve in the same figure shows that $(F/F_0)\sigma^{-0.6}$ is almost constant, indicating that F is roughly proportional to $\sigma^{0.6}$, as shown also for a civil turbofan in Figure 5.9.

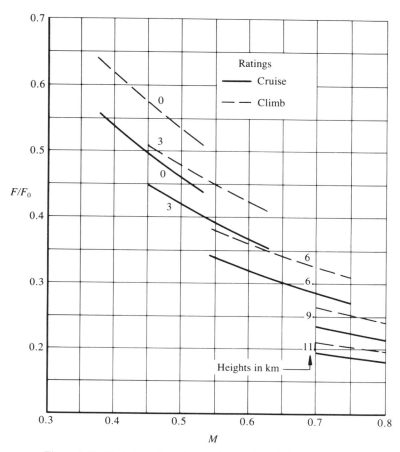

Figure 5.25. Variation of maximum thrust with height and Mach number for propfan 'C'.

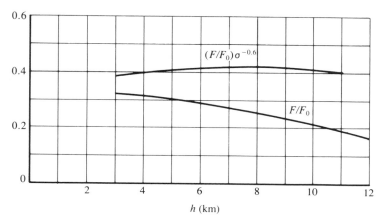

Figure 5.26. Variation of maximum cruise thrust with height at $M = 0.7$ for propfan 'C'.

5.5.2 FUEL CONSUMPTION

Defining specific fuel consumption as $c = Q/F$, as for a turbofan, it is found that the sfc of a propfan increases with flight speed at a fixed height. This is shown for the propfan 'C' at the cruise rating in Figure 5.27, where the notation is the same as in Figure 5.10, i.e. c_{so} is the value of c in static conditions at the take-off rating at sea level. As for a civil turbofan, the variation of c with M is well represented by Equation (5.11), with suitably chosen values of c_1 and n. The broken lines in Figure 5.27 show that for heights of 6 and 11 km good agreement with the measured data is obtained with $n = 0.85$, but for lower heights it is found that $n = 0.77$ gives better agreement. The required values of n may of course be somewhat different for other propfans.

Reference was made in § 5.4.2 to the results of a survey given in ESDU 75018 (1975), showing that Equation (5.11) could be used with $n = 0.9$ to give c (not c') for a turboprop, and this was approximately confirmed by the results given in Figure 5.20 for the small turboprop 'B'. The results now given show that for the propfan 'C' the form of variation of c with M at a fixed height is not much different from that found for turboprops.

Figure 5.27 shows that for the propfan 'C' at a fixed Mach number there is some decrease of c with increasing height. It was shown earlier in discussion of Figure 5.11 that for a civil turbofan the change of c with height was given with reasonable accuracy by Equation (5.11), but further analysis of the results shown in Figure 5.27 shows that this is not true for the propfan 'C'. Whereas Equation (5.11) gives $c \propto \theta^{1/2}$ it is found that $c \propto \theta$ represents the propfan 'C' more closely, although even with this law there are discrepancies of about ±3%.

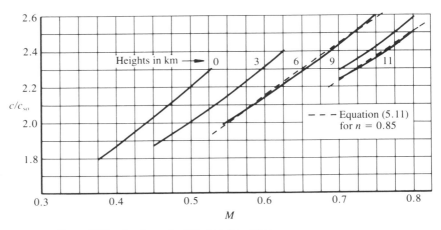

Figure 5.27. sfc of propfan 'C' at cruise rating.

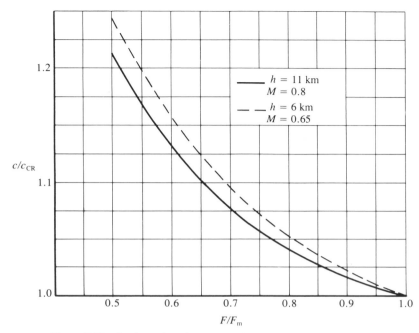

Figure 5.28. sfc of propfan 'C' at reduced thrust.

If the throttle control is used to reduce the thrust below the rated cruise value there is an increase of sfc and this is shown for the propfan 'C' in Figure 5.28. The notation used here is the same as in Figure 5.12, i.e. c_{CR} and F_m are respectively the sfc and the thrust at the rated cruise condition. The increase of c shown in Figure 5.28 is considerably greater than that shown for two civil turbofans in Figure 5.12 and a comparison can also be made with the data for the turboprop 'B' shown in Figure 5.22, even though the quantities plotted in Figures 5.22 and 5.28 are different. If it can be assumed that for the turboprop P_e/P_s and η_{PR} are constant, then $F/F_m = P_s/P_{sm}$ at a fixed speed and Equation (5.23) shows that $c/c_{CR} = c_s'/c_{sCR}'$. With these assumptions Figures 5.22 and 5.28 can be compared and it is apparent that the increase of sfc shown in Figure 5.28 is of the same order as that shown in Figure 5.22 for the turboprop 'B' at the lower altitudes.

5.6 Piston engines

As mentioned earlier, piston engines driving propellers are now used only on small aircraft with powers less than about 300 kW. On some very simple light aircraft the propeller may be of fixed pitch;

increasing flight speed then gives an increase of rotational speed and a consequent increase of maximum shaft power. In the more usual case where the propeller is of variable pitch and is controlled to keep the rotational speed constant, increase of flight speed has no appreciable effect on the maximum shaft power, since aircraft using piston engines fly only at low Mach numbers and hence the effects of intake ram pressure are negligible. For these aircraft, as for those using turboprops, a knowledge of propeller efficiency as a function of J is required for calculation of aircraft performance, so that the thrust F can be found from the shaft power for any given forward speed.

If the engine is normally aspirated (i.e. not supercharged) the reduction of air density with increasing height causes a decrease of maximum power for any given rotational speed. Torenbeek (1982) represents this by the empirical equation

$$P_s/P_{so} = (1 + c_h)\sigma - c_h \tag{5.26}$$

and suggests a value of 0.132 for the constant c_h, although this is based only on limited experimental evidence for several radial engines. In a supercharged engine a compressor is used to increase the pressure in the inlet manifold, so that the power at high altitudes can be greatly increased. The compressor may be driven either through gearing from the crankshaft or by a small turbine using the energy in the engine exhaust (turbocharging). In either case the inlet manifold pressure is usually limited by throttling to a constant preset value at low altitudes, to avoid overstressing the engine. This constant pressure is maintained up to the *critical altitude* or *full-throttle height* and below this height the maximum shaft power for a given rotational speed increases slightly with increasing height, because of the reduction of air temperature and consequent increase of density for a given pressure. Above the critical altitude the maximum power decreases with increasing height as for a normally aspirated engine.

For a piston engine the specific fuel consumption $c_s' = Q/P_s$ depends on the fuel/air ratio, but at the optimum value of this ratio and at maximum cruise power c_s' is nearly independent of height and forward speed. There is often some increase of c_s' as power is reduced to a low level but this is usually less severe than for a turboprop engine.

5.7 Summary of conclusions

A summary is given below of the more important conclusions from the data presented in this chapter, concentrating on the variations with height and speed of the maximum available thrust or power and the specific fuel consumption. The conclusions are based on data obtained from only a few engines and may not apply to all engines in the class considered, e.g. military turbofans or propfans.

Summary of conclusions

5.7.1 MAXIMUM THRUST OF TURBOJETS, TURBOFANS AND PROPFANS

In the ISA at heights between 11 and 20 km the air temperature is constant and the functional relations (5.5) and (5.7) show that F/δ and c depend only on the Mach number, if the engine speed is constant. For lower heights these two quantities depend both on Mach number and on height.

In the range of speeds used for take-off Equation (5.8) may be used, with suitably chosen values of k_2 and k_3, to represent the variation of the maximum thrust F_m with forward speed V. For a propfan the constant k_3 may sometimes be assumed to be zero, with little loss of accuracy.

For civil turbofans in the range of speeds used for subsonic cruising, the variation of F_m with V at a fixed height can usually be neglected, but for a civil turbofan in a climb, or for a propfan either cruising or climbing, Equation (5.9) should be used, giving $F_m \propto M^{-n}$. For a civil turbofan the exponent n is likely to be in the range 0.2–0.4, but for a propfan it is much greater, typically about 0.6–0.7. For a typical military turbofan there is a substantial increase of maximum thrust with Mach number, into the supersonic speed range, and a power law such as Equation (5.9) is not useful, with any value of the exponent n.

For all turbojets, turbofans and propfans there is a substantial loss of F_m as the height increases at constant Mach number. For $h > 11$ km in the ISA this is given by $F_m \propto \sigma$ and for a civil turbofan or a propfan at lower heights there is evidence that a good approximation is given by $F_m \propto \sigma^{0.6}$.

5.7.2 sfc OF TURBOJETS, TURBOFANS AND PROPFANS

For a dry military turbofan, or a turbojet as used for supersonic flight, the specific fuel consumption c shows only a modest increase with Mach number at subsonic and moderate supersonic speeds, but at higher supersonic speeds there is a rapid increase of c with M. For a civil turbofan or a propfan (at subsonic speeds) there is an increase of sfc with Mach number as given by Equation (5.11). The exponent n in this equation is typically about 0.5 for a civil turbofan and in the range 0.7–0.9 for a propfan. For a civil turbofan the same equation may be used to represent the variation of c with height as well as Mach number, but for a military turbofan there is little variation of c with height, whereas for a typical propfan the reduction of c with increasing height is greater than that given by Equation (5.11).

For a military turbofan a reduction of thrust below the cruise rated value (say by 30%) gives a reduction of c (up to about 10%), but further reduction of thrust leads to an increase of c. For a civil turbofan the reduction of c is smaller and may not always occur, whereas for a typical propfan there is no reduction of c and any reduction of thrust gives an increase.

5.7.3 TURBOPROPS

For a turboprop in the range of speeds used for climbing or cruising the maximum shaft power at a fixed height is given by Equation (5.20) as $P_s = P_{so} A M^n$. The exponent n will always lie between 0 and 1 and in a typical case might be about 0.5. The variation of shaft power with height, at a constant Mach number, can be represented by $P_s \propto \sigma^n$, but the value of n in this relation must be determined from test data for each particular engine.

The sfc of a turboprop, expressed as $c' = Q/P_e$, usually varies only slightly with height. The variation with M is also fairly small and may often be neglected, but it can be represented more accurately by $c' \propto M^{-m}$, where m usually lies in the range from 0 to 0.2. As the power is reduced below the cruise rated value there is an increase of c', similar to the increase of c noted earlier for a propfan.

6

Take-off and landing performance

In most of the traditional areas of performance studies the central effort is directed towards a refinement of aerodynamic design in order to ensure efficiency of flight. In studying take-off and landing performance attention is directed more to the capacity of the engines to accelerate the aircraft in a condition of high drag coefficient when the available distance is limited, and to the braking capacity during landing for the same reason. There is limited opportunity for refinement of aerodynamics, although in recent years some progress has been made in reducing the drag of an aircraft in the take-off configuration. In the crucial periods close to lift-off and touchdown the behaviour of the aircraft is strongly dependent on piloting technique and there is a need to define standard procedures for these two manoeuvres, based on reference speeds which are used in defining the criteria laid down in airworthiness regulations for safe operations. The lengths of the ground run and the airborne sector in a take-off are both strongly influenced by engine performance and the effective drag polar, whereas in landing there is an airborne sector which is critically dependent on piloting technique to produce a tangential flare to touchdown, followed by a ground run which depends on braking capacity.

The take-off performance cannot be defined so simply when allowance is made for failure of an engine and this chapter considers not only the ideal performance when the manoeuvre proceeds as planned, but also the reduced performance obtained after an engine failure. Consideration is also given to take-off and landing performance when the air pressure and temperature are significantly different from the values in the ISA at sea level and when there is a substantial component of wind velocity along the runway.

A few unusual classes of aircraft are excluded from this chapter by some of the assumptions stated in Chapter 1, viz. that the propulsive thrust acts in the direction of motion of the aircraft and there is no

120   Take-off and landing performance

interaction between this propulsive thrust and the aerodynamic force. Thus aircraft with powered lift systems such as blown flaps and jet flaps are excluded unless the air supply to the lift system is controlled independently of the propulsive thrust. Aircraft with provision for varying the direction of the thrust line (vectored thrust) are also excluded from this chapter but will be discussed in Chapter 9.

Both of the manoeuvres considered here, take-off and landing, normally employ the special benefits of flaps and other high-lift devices, although these contribute to the high drag mentioned earlier. The extension of the undercarriage increases the drag still further, but proximity of the wing to the ground usually gives a reduction of drag. It is towards these special influences on lift and drag that attention is turned first, before considering the two manoeuvres later in some detail.

6.1   High-lift devices

A high maximum lift coefficient C_{Lmax} is desirable, both for take-off and for landing, because this reduces the stalling speed for a given wing loading and so reduces the speeds needed in the airborne phase and hence reduces the length of the required ground run. Alternatively, if the length of the ground run is fixed, an increase of C_{Lmax} alllows an aircraft to take-off or land at a greater total weight (unless the weight is limited by the structural design), provided that the thrust is still sufficient to accelerate the heavier aircraft to the take-off speed in the available distance and that the braking capacity is sufficient to stop the aircraft within the specified length.

The usual method of increasing the maximum lift coefficient of a wing for take-off or landing is by the use of slats and flaps and the sketches (*a*) to (*c*) in Figure 6.1 show a commonly used arrangement with a slat at the leading edge and a double slotted flap at the trailing edge. In the cruising configuration the slat and flap elements are all retracted into the wing as in the basic shape (*a*). Progressive extension of the slat and flap as shown in (*b*) and (*c*) gives a substantial increase of C_{Lmax} because of the high overall camber and the favourable interaction between the multiple elements of the aerofoil as explained by Smith (1975). For take-off, a very high drag cannot be tolerated because this would increase the length of runway required and would not allow a large enough angle of climb to be maintained after failure of one engine. For this reason a configuration like that in (*b*) is normally used for take-off, with the angle δ_f between the chords of the rear flap and the main aerofoil no more than about 15° or 20°, as shown in Figure 6.2. For landing, a very high drag is acceptable and may even be advantageous because it allows a descent to be made at a steep angle. Thus for landing, a configuration as in (*c*) is normally

High-lift devices

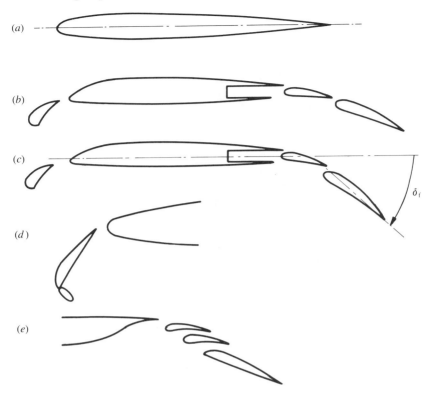

Figure 6.1. High-lift devices.

Figure 6.2. Boeing 747 just after lift-off. The leading-edge and trailing-edge flaps can be seen at the setting used for take-off.

used with a flap deflection angle δ_f of about 40° or more, as shown in Figure 6.3. This gives a higher value of C_{Lmax} than configuration (b) and a considerably higher drag.

Although the arrangement shown in (a) to (c) of Figure 6.1 is often used there are other possibilities. A nose flap or drooped leading edge may be used, either instead of a slat or in a form that incorporates a slot as in the outer part of the Boeing 747 wing described by McIntosh & Wimpress (1975) and shown in (d). In other designs the flap at the trailing edge may have only a single slot for greater simplicity or it may be of *triple slotted* form as shown in (e). It should be noted that although the extension of a flap at the trailing edge usually also gives a significant increase of lifting area, it is customary to base all the coefficients of lift and drag on the area of the 'clean' wing with the slat and flap retracted.

A useful survey of high-lift flaps and slats and methods of predicting their effects on a wing has been given by Callaghan (1974). As shown by the curves A and B in Figure 6.4 the principal effect of a slat at the leading edge is to delay stalling to a much higher lift coefficient and angle of incidence, with little effect on C_L at a given α. Two-dimensional experimental data given by Callaghan show that a slat increases C_D by an amount of order 0.01, but of course this can apply only to a wing on which separation does not occur near the leading edge with the slat retracted. In cases where extension of the slat suppresses separation this is likely to *reduce* the overall value of C_D. As shown by the curves B and C in Figure 6.4 the two principal effects of a trailing edge flap on lift are an increase $(\Delta C_L)_\alpha$ in the lift coefficient at a constant (low) incidence and an increase ΔC_{Lmax} in the

Figure 6.3. Boeing 747 just after touchdown. In this landing configuration there is a large deflection of the trailing-edge flaps.

High-lift devices

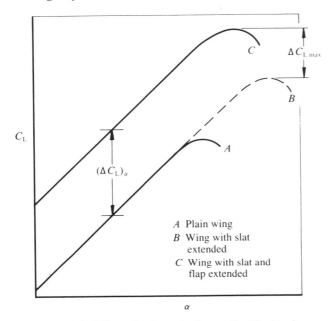

Figure 6.4. Effects of a slat and a flap on the lift of a wing.

maximum lift coefficient. The latter quantity ΔC_{Lmax} is almost always less than $(\Delta C_L)_\alpha$. The increment $(\Delta C_L)_\alpha$ depends mainly on the flap angle δ_f and on the ratio of flap chord to main wing chord. This increment can be estimated theoretically with good accuracy, especially for flap angles that are not very large, but the increment ΔC_{Lmax} is much more difficult to estimate theoretically and would normally be obtained from wind tunnel tests at a high Reynolds number and appropriate Mach number.

For an aircraft with high-lift flaps extended it is possible to estimate theoretically the spanwise lift distribution and hence calculate the vortex drag coefficient with reasonable accuracy. There is much greater difficulty in estimating the component of drag coefficient due to viscous effects, because at high C_L the boundary layer on much of the wing and on the flap is thick and may be close to separation. Also, for each element of the system, e.g. the four elements of Figure 6.1(c), there may be a complex interaction between the boundary layer on that element and the wake of an upstream element (slat, wing or forward flap). Moreover the flap operating mechanism is exposed to the air stream when the flaps are extended and this necessarily includes many parts whose drag is difficult to estimate. Thus wind tunnel tests are the only reliable source of data on this component of drag.

A further component of drag caused by extension of high-lift flaps is the *trim drag*, the additional drag introduced by the need to adjust the tailplane and elevator angles so that the overall pitching moment about the centre of gravity of the aircraft is zero. Extension of the flap at the trailing edge causes a rearward shift of the centre of pressure on the wing and the resulting pitching moment must be counteracted by a downward load (or smaller upward load) on the tailplane. This requires an increase of wing lift to maintain a given overall lift, so that there is an increase of lift-dependent drag on the wing. There is also usually an increase of tailplane drag and Callaghan (1974) has shown that the overall trim drag may be as much as 5% of the total drag of the aircraft.

6.2 Drag of the undercarriage

ESDU 79015 (1987) gives an empirical equation for the drag of an undercarriage for use at the feasibility study stage of a project when the dimensions of the undercarriage are not yet known. This is based on an equation given by Torenbeek (1982) and is derived from data on a number of civil transport aircraft. The equation shows that the increase of drag coefficient due to the undercarriage is approximately

$$\Delta C_D = w K_{uc} m^{-0.215}, \tag{6.1}$$

where m is the maximum mass of the aircraft at take-off in kg and w is the corresponding wing loading in N/m². The value of the coefficient K_{uc} is given as 5.81×10^{-5} for zero flap deflection and 3.16×10^{-5} for maximum flap deflection. For $w = 5000$ N/m² and $m = 10^5$ kg these figures give values of ΔC_D of 0.024 and 0.013 for zero and maximum flap deflection respectively. The decrease of undercarriage drag with increasing flap deflection is supported by flight test data for a DC-8 given by Callaghan (1974) and by data for the Boeing 747 given by Hanke & Nordwall (1970), which also show a reduction of ΔC_D with increasing angle of incidence. These effects are apparently caused by the reduction of relative air velocity at the undercarriage, for a given flight speed, as the flap angle or incidence increases.

ESDU 79015 (1987) also gives a method of calculating the drag of an undercarriage, when the dimensions of all the components are known, by summing the drags of the individual components and introducing empirical factors to allow for the effects of interference among the components of the undercarriage and also interference between the whole undercarriage and the wing. One of these factors allows for the effects of varying flap deflection, but in contrast to the results mentioned earlier for the Boeing 747 it is found that there are no significant effects of varying lift coefficient (at constant flap

Effects of ground proximity on lift and drag 125

deflection), apparently because in most cases the mutual interference effects between the wing and the undercarriage roughly counteract one another.

6.3 Effects of ground proximity on lift and drag

Proximity of the ground has important effects on the relationship between C_L, C_D and the angle of incidence α. These effects are discussed by Torenbeek (1982) and further information is given in ESDU 72023 (1972). For the purpose of a mathematical model the ground is replaced by an image of the bound and trailing vortex system of the aircraft, so that the condition of zero vertical velocity component at the ground is satisfied.

The image of the bound vortex representing the wing of the aircraft reduces the effective stream velocity at the wing, thus reducing the lift for a given α, but it also increases the effective camber and incidence of the wing, tending to increase the lift. These opposing effects roughly cancel one another except at very low heights above the ground and with high lift coefficients, where the effect of the reduced stream velocity is dominant and there is a loss of lift.

In contrast to the effect of the bound vortex, the images of the trailing vortices induce an upward velocity component (upwash) at the wing and this gives an increase of lift. Results of calculations given by Torenbeek (1982) show that for lift coefficients up to about 2 the net effect of the whole image system is an increase of lift for a given α, but for higher lift coefficients and very low heights the lift is reduced because of the dominant effect of the reduced flow velocity. Torenbeek (1982) remarks that distortion of the pressure distribution on the wing caused by the ground usually leads to a reduction in the maximum lift coefficient, especially when trailing edge flaps are extended.

Proximity of the ground causes a reduction of drag for two reasons:

(a) the upwash induced at the wing by the images of the trailing vortices partially cancels the downwash due to the real trailing vortices, with the consequence that the rearward inclination of the lift vector is reduced and thus the normal vortex drag is partially suppressed,

(b) the reduction of effective stream velocity caused by the image of the bound vortex reduces the viscous drag for a given value of C_D, since this is proportional to $\frac{1}{2}\rho V^2$.

The reduction of drag caused by this 'ground effect' becomes important for small heights and this may be illustrated by an example. For a typical aircraft with a wing of aspect ratio 7 and $C_L = 1.6$, with the wing at a height above the ground equal to $\frac{1}{5}$th of the span, the equations given in Section G-7 of Torenbeek (1982) show that the reduction of vortex drag due to the effect (a) is about 24% and the reduction of viscous drag due to (b) is about 6%. Further calculations show that with the undercarriage down the reduction of total drag might then be about 16% with the flaps set for take-off, or about 13% in the landing configuration when the flap angle is greater.

6.4 Drag equations for take-off and landing

As an example to illustrate the validity of the simple parabolic drag law, even for an aircraft in the take-off or landing configuration, Figure 6.5 shows some results derived from data given in Section 11.3 of Torenbeek (1982) for a short-haul airliner. The results refer to the aircraft with slats and flaps extended and with the undercarriage retracted. With the undercarriage down Equation (6.1) suggests that C_D would be increased at all values of C_L by about 0.022 with the flaps at 15° and by about 0.015 with the flaps at 45°. The broken straight lines in Figure 6.5 show that in this case the drag polar is well represented by the simple parabolic drag law for $C_L \leq 1.7$ with the flaps at 15° and for $1.25 \leq C_L \leq 2$ with the flaps at 45°. The values of K_2 given by the slopes of the straight lines are very nearly the same for the two flap angles and are also close to the value obtained from

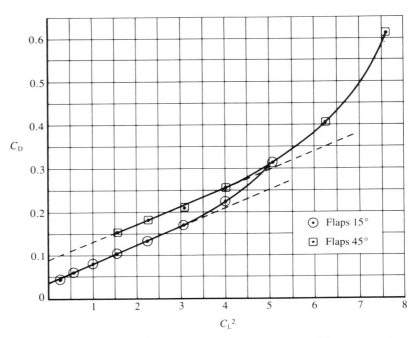

Figure 6.5. $C_D - C_L^2$ relation for an aircraft with slats and flaps extended, as given by Torenbeek (1982).

Torenbeek's data for the aircraft with slats and flaps retracted. Figure 6.5 shows that if K_1 and K_2 in Equation (3.6) are chosen to fit the data in the lower range of C_L the equation will underestimate C_D at the higher values of C_L. This probably occurs because of boundary layer separation on the flap at the higher values of C_L, leading to greatly increased viscous drag so that Equation (3.4) is no longer even approximately correct. It is possible to develop equations to represent with fair accuracy the true variation of C_D in the higher ranges of C_L, but for the purpose of calculating take-off and landing performance this is often not necessary because the lift coefficients are low enough to lie within the range for which C_D varies linearly with C_L^2.

The increased drag caused by lowering the undercarriage and extending the flaps leads to a large increase in K_1 in Equation (3.6). Taking as an example the aircraft represented in Figure 6.5, K_1 is about 0.018 in the 'clean' configuration with undercarriage up and flaps retracted. With undercarriage down and flaps extended K_1 is increased to about 0.06 at 15° flap angle and 0.10 at 45°. There is little change of K_2 and as a result $C_L{}^*$ increases from about 0.62 in the clean condition to about 1.2 at 15° flap angle and 1.6 at 45°, giving a reduction of both the speeds $V_e{}^*$ and V_{ec} discussed in Chapter 2. If there were no change of stalling speed this would make it less likely that V_e would be lower than V_{ec} in any airborne part of the take-off or landing, but of course extension of the slats and flaps does give a large reduction of stalling speed and this has the opposite effect, making it more likely that V_e will be less than V_{ec}.

6.5 Take-off procedure and reference speeds

Figure 6.6 shows the path of an aircraft during take-off and some of the most important speeds associated with successful completion of the manoeuvre. The aircraft starts from rest and accelerates along the ground at a nearly constant angle of incidence which is determined by the design of the undercarriage and by the distribution of the aircraft weight upon it. When the *rotation speed* V_R is reached the pilot uses the elevator control to begin the rotation of the aircraft in the nose-up sense, usually achieving a mean rate of rotation in the

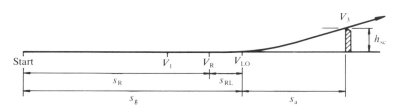

Figure 6.6. Take-off path (not to scale).

region of 3° per second for a large transport aircraft, but greater for some smaller transports and military aircraft. As rotation begins the angle of incidence and lift coefficient increase and soon after the start of rotation the nose wheel leaves the ground. As the lift becomes equal to and then exceeds the aircraft weight the main wheels leave the ground, the speed at which this occurs being the *lift-off speed* V_{LO}. After lift-off there is a transition phase in which the flight path is curved and the lift must be greater than the weight, followed by a steady climb at an angle γ_c. In order to ensure safe operation in realistic conditions the top of a hypothetical screen (or standard obstruction) of height h_{sc} must be cleared before the take-off is considered complete, this occurring at a speed V_3 which must not be less than the *take-off safety speed* V_2. The screen height may be reached either before or after the end of the curved transition phase.

Requirements for certification of airworthiness are issued by the Joint Airworthiness Authority (JAA) in Europe and by the Federal Aviation Administration (FAA) in the USA. These requirements define and employ a number of reference speeds, few of which have definite values which can be calculated uniquely even when a specific aircraft configuration and the local environmental conditions are known. Most of the speeds are defined as minimum multiples of the few values which can be calculated and indeed some speeds must satisfy more than one such criterion. The most important definitions and relationships are outlined below.

(a) The *stalling speed* V_s is defined as the minimum speed at which level flight can be maintained with zero acceleration, usually with the minimum thrust necessary to maintain this condition. This is an important reference speed and can be calculated if the maximum lift coefficient is known.

(b) The *minimum unstick speed* V_{MU} is the lowest speed at which the aircraft can be made to leave the ground and continue to climb. In some cases this speed may depend on the maximum rotation angle that can be obtained before the rear fuselage touches the ground and as this angle may give a lift coefficient that is less than C_{Lmax} the associated speed may exceed V_s by a small amount. This too is a speed used for reference and is one that can be calculated for known conditions.

(c) There are two *minimum control speeds*, V_{MCg} for movements while the aircraft is still on the ground and V_{MCa} for airborne flight, the latter being the more important reference speed. Both speeds are related to directional control after an engine failure and the ability of the pilot to recover from the sudden yaw disturbance it creates. On the ground, recovery is easier than in the air, but the regulations do not allow the advantageous use of nose-wheel steering and hence there is

Take-off procedure and reference speeds

still a requirement for strong counteraction by aerodynamic control force from the fin and rudder. Nevertheless it is clear that V_{MCg} will be somewhat less than the crucially important speed V_{MCa}. Control or recovery at the speed V_{MCa} must be possible while the aircraft is airborne with one engine failed and with maximum thrust from the remaining engines. The speed is determined by the ability of the fin and rudder to develop a yawing moment sufficient to counteract the effect of the asymmetric thrust from the engines and is therefore of prime importance in determining the required sizes of the fin and rudder. These surfaces must be large enough to provide the required balancing yawing moment at a speed V_{MCa} which does not exceed 1.2 V_s at the maximum take-off weight.

(d) The *decision speed* V_1 is necessarily less than V_R and has a special significance in relation to the possibility of an engine failure during the take-off. If the speed V_{EFR} at which an engine failure is recognised is less than V_1 the take-off must be abandoned and the aircraft brought to rest, whereas if $V_{EFR} > V_1$ the take-off must be continued. The speed V_1 and the consequences of engine failure will be discussed more fully in § 6.6.

(e) The speed V_R at which rotation is initiated must be at least 1.05 V_{MCa} and must also be high enough to ensure that, after rotation at the maximum possible rate, the lift-off speed V_{LO} will not be less than 1.1 V_{MU} with all engines operating or 1.05 V_{MU} after failure of one engine. To satisfy the latter condition with all engines operating, V_R is usually required to be not less than 1.1 V_{MU}. (The increase of speed during the rotation phase is small, so that V_{LO} is usually only a little greater than V_R.)

(f) The most important target speed in the take-off manoeuvre is the take-off safety speed V_2 which is specified to ensure an adequate safe climb-out after one engine has failed, the requirement being that the aircraft should be able to achieve this speed no later than the point at which the screen height is reached. (In many cases the maximum take-off weight will be limited by the need to achieve V_2 and the desired climb angle.) Clearly there is an implication of continued acceleration after lift-off, even with one engine failed, and the minimum requirement for V_2, before considering the climb criterion, is that it should not be less than either 1.2 V_s or 1.1 V_{MCa}. With *all* the engines operating the speed V_3 obtained at the screen height will be only moderately greater than V_2, typically about 5 m/s greater when V_2 is of order 80 m/s.

A more extensive account of these reference speeds is given by Williams (1972) and the requirements are specified in detail in the European Joint Airworthiness Requirements (JAR)-25 and in the US Federal Aviation Regulations (FAR) Part 25. In order to clarify the criteria specifying the relationships between the reference speeds, Figure 6.7 displays a résumé of the multiplying factors relating the speed shown at the head of each arrow to that shown at its tail. A plus sign indicates that the factor must be at least as great as the value shown, whereas a minus sign means that the factor must be no greater than the value shown. For example the arrow from V_{MU} to V_R means that V_R must not be less than $1.1 V_{MU}$. Figure 6.7 illustrates the fact that only V_s, V_{MU}, V_{MCg} and V_{MCa} can be determined uniquely.

6.6 The balanced field length and the take-off transition

The decision speed V_1 is determined by considering the two possible alternative actions after recognition of an engine failure at a speed V_{EFR},

(a) continuing the take-off,
(b) keeping the aircraft on the ground and bringing it to a stop.

(The speed V_{EFR} is greater than the speed V_{EF} at which the engine failure actually occurs, because there is a short delay period before the pilot recognises the failure.) As V_{EFR} increases towards the rotation speed V_R, the distance to clear the screen height for (a) decreases whereas the distance for (b) increases. The decision speed V_1 is chosen as the value of V_{EFR} for which these two distances are equal. Thus the

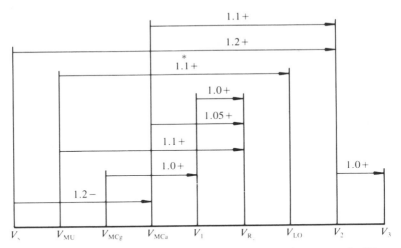

Figure 6.7. Multiplying factors relating take-off reference speeds. (No sense of scale is implied.)

In some cases the requirements differ slightly between different classes of aircraft, e.g. between light aircraft and transport aircraft. They may also depend on whether one engine has failed or not, e.g. the factor marked * relating V_{MU} to V_{LO} can be as low as 1.05 after failure of one engine on a transport aircraft.

total distance required for (a) continues to decrease as V_{EFR} increases above V_1, whereas the distance required for (b) decreases as V_{EFR} becomes progressively smaller than V_1. When $V_{EFR} = V_1$ the total distance required for either of the actions (a) or (b) is the same and is known as the *balanced field length*. The allowable stopping distance for (b) includes a short 'stopway' beyond the end of the normal runway and under the conditions which define the balanced field length there is an implication that the hypothetical screen is at the far end of this stopway, thus making the two distances equal. There is an overriding requirement that V_1 must be greater than the minimum control speed on the ground V_{MCg}, since if this condition were not satisfied the action (a) would not be possible after recognition of an engine failure at speed V_1.

Figure 6.8 shows the distinction between (i) the normal take-off performance with all engines operating and (ii) the performance after an engine failure in the worst case with $V_{EFR} = V_1$. In the trajectory for (i), marked ae, the screen height is reached at a distance Δs before the screen and the speed V_3 at this height is greater than V_2. The trajectory marked fe refers to the case where failure of one engine is recognised at the speed V_1 and the take-off is continued. The speeds for rotation and for lift-off are reached later in the ground run and the speed V_2 is just reached at the top of the screen. The distance shown as s from the decision point at speed V_1 to the screen is the same as the distance that would be required from the decision point to bring the aircraft to rest on the ground. As V_{EFR} increases above V_1 the failed engine trajectory moves to the left on the diagram and the screen height is reached at an increasing distance before the screen, this distance becoming equal to Δs when the engine failure occurs only when the screen height is reached. Similarly, as V_{EFR} falls below V_1 the point at which the

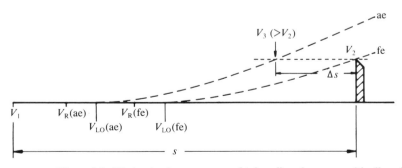

Figure 6.8. Distinction between normal take-off performance with all engines operating (ae) and minimum performance after recognition of the failure of one engine (fe) at speed V_1 (not to scale).

aircraft can be stopped moves progressively farther ahead of the screen.

For the purpose of airworthiness certification it is usual to divide the airborne part of the take-off and the subsequent climb into three segments. In the first segment, shortly after lift-off, the undercarriage and flaps are still extended, giving an increase in both the lift and the drag, but any favourable effects of ground proximity on lift and drag are not taken into account. In the second segment, extending to a height of about 120 m, the undercarriage is assumed to have been retracted but the flaps and slats are still at the take-off setting, while in the third segment, extending to a height of about 450 m, different lesser settings of the flaps and slats may be used, thus progressively reducing their drag penalty to allow the speed to build up as height is gained. For each of the three segments a minimum angle of climb after an engine failure is specified, the angle for the second segment usually giving the critical design case. For this segment the specified angles in radians are 0.024 for a two-engined aircraft, 0.027 for three engines and 0.030 for four.

In Figure 6.6 the length s_R is the distance required to accelerate from rest to the rotation speed V_R. The additional length s_{RL} required to reach the lift-off condition is difficult to estimate accurately but this length is usually no more than about 200 m for a typical transport aircraft, so that errors in its estimation have relatively small effects on the total length of ground run s_g. The distance s_a is the length of the airborne part of the take-off, from lift-off to achievement of the screen height, and the total length required for take-off is $(s_g + s_a)$.

Before proceeding to an estimation of these take-off lengths it should be noted that an accurate calculation would involve at least the following, just for the sector for which $V > V_1$:

1. Establish an adequate gradient for the steady climb above the screen height, perhaps allowing for a specific distant obstruction.
2. Set a value for V_2 to support flight at this gradient for the known aircraft weight, allowing for engine failure.
3. Find a value for V_{LO} which is consistent with estimates of acceleration in the curved transition sector and determine the sector length from lift-off to the screen height.
4. Use estimates of acceleration and rotation time to establish values for V_R and the length s_{RL} shown in Figure 6.6.
5. Calculate back from V_R to V_1 by using a standard delay time and an acceleration rate that takes account of the lift and drag coefficients in the take-off configuration close to the ground. Hence the runway length covered in the acceleration from V_1 to V_R can be estimated.

The complete procedure will necessarily be more complex and must be an iterative one to ensure the satisfaction of all the criteria indicated in Figure 6.7 and discussed earlier. The procedure which is outlined below is considerably simpler and is based on an assumption that the reference speeds can be readily estimated.

6.7 The take-off ground run

It is instructive to consider first a very simple approximate calculation in which the thrust is assumed to be constant and the rolling resistance and aerodynamic drag are neglected. Then consideration of energy shows that the distance required to accelerate on a level runway from rest to a speed V is

$$s = \tfrac{1}{2}mV^2/F. \tag{6.2}$$

If C_L is the lift coefficient for steady level flight at the speed V,

$$V^2 = 2W/(\rho_0 \sigma S C_L) \tag{6.3}$$

and so

$$s = W^2/(Fg\rho_0\sigma S C_L) = w/(fg\rho_0\sigma C_L), \tag{6.4}$$

where $w = W/S$ and $f = F/W$. Putting $V = V_{LO}$ in Equation (6.3) and using an appropriate mean value for the thrust F, Equation (6.4) gives a value of s that is less than the true distance s_g as shown in Figure 6.6, because the effects of rolling resistance and aerodynamic drag have been neglected. Nevertheless the quantity $w/(f\sigma C_L)$ is useful as a take-off parameter, because measured values of s_g for a number of aircraft can be used to derive an empirical factor which when multiplied by that parameter gives an approximate estimate of s_g.

The rolling resistance of an aircraft on the ground is usually assumed to be proportional to $(W - L)$, the net downward load exerted by the aircraft on the undercarriage. Thus the rolling resistance is expressed as $\mu_R(W - L)$, where μ_R is a constant coefficient of rolling resistance and is usually taken to be about 0.025 for a concrete runway. Collingbourne (1970) has shown that μ_R depends to some extent on the forward speed and on the inflation pressure and other characteristics of the tyres, but the exact value of μ_R that is used in estimating the length of the ground run is not very important because the rolling resistance is usually small in comparison with the thrust. An important exceptional case occurs when the runway is covered with snow or slush, giving a large increase of μ_R which has a substantial effect on the length of the ground run.

For an aircraft with the usual form of nose-wheel undercarriage the angle of incidence may be assumed to be constant at all speeds below V_R, since any change due to differential extension of the undercarriage legs as the lift increases will be very small. Variations of C_L and C_D due to changes of Reynolds number may also be neglected, because the only significant changes will be in C_D at very low speeds, where the total aerodynamic drag force is in any case negligible. Thus C_L and C_D may be assumed to have the constant values C_{LG} and C_{DG} at all speeds up to V_R, although they both change substantially in the rotation phase between the speeds V_R and V_{LO}.

The equation of motion for an aircraft accelerating on a runway with a small upward gradient γ_G is

$$m\,dV/dt = mV\,dV/ds = F - D - \mu_R(W - L) - \gamma_G W$$
$$= F - \tfrac{1}{2}\rho V^2 S(C_D - \mu_R C_L) - (\mu_R + \gamma_G)W, \qquad (6.5)$$

but in the remainder of this chapter it will be assumed that the runway is level and the term $\gamma_G W$ will be omitted. The acceleration decreases as the speed rises, partly because of the V^2 term in the equation and partly because of the reduction of thrust that occurs as the speed increases, especially for turboprops and propfans and for turbofans with high by-pass ratios.

In order to simplify the calculation a mean acceleration $\bar{a}$ is sometimes defined for the ground run up to the speed V_R and the energy approach used in deriving Equation (6.2) is then used to find an approximate value of the distance s_R. For turbofan engines the mean acceleration $\bar{a}$ is calculated by putting $V = 0.7 V_R$ in Equation (6.5) and by putting F in the same equation equal to the thrust at this speed. Since the distance s_R is required the kinetic energy should of course be the value at the full rotation speed V_R and the approximate estimate of the distance is given by the equation for uniform acceleration

$$s_R = \tfrac{1}{2}V_R^2/\bar{a}. \qquad (6.6)$$

For propeller driven aircraft Dekker & Lean (1962) suggest that the mean acceleration $\bar{a}$ should be calculated for a speed $V = 0.74 V_R$ and this value may also be used for propfans, since these have thrust-speed characteristics similar to those of turboprops.

It is sometimes assumed that the variation of thrust with speed is given by

$$F/F_0 = 1 - k_1 V^2, \qquad (6.7)$$

where k_1 is a positive constant, although ESDU 76034 (1976) and the data given in Chapter 5 show that this is not a good representation, either for turbofans, turboprops or propfans. Nevertheless the calculation of distance from Equation (6.5) can be greatly simplified by assuming that F depends only on V^2 and if Equation (6.7) is assumed

The take-off ground run

to be valid the distance s_R is obtained as

$$s_R = \frac{W}{2g} \int_0^{V_R} \frac{d(V^2)}{F_0 - k_1 F_0 V^2 - \frac{1}{2}\rho V^2 S(C_{DG} - \mu_R C_{LG}) - \mu_R W}, \quad (6.8)$$

where C_{DG} and C_{LG} refer to the aircraft on the ground, allowing for ground effect. If a constant is introduced as

$$K_G = C_{DG} - \mu_R C_{LG} + k_1 F_0 / (\tfrac{1}{2}\rho S), \quad (6.9)$$

the Equation (6.8) takes the simpler form

$$s_R = \frac{W}{2g} \int_0^{V_R} \frac{d(V^2)}{W(f_0 - \mu_R) - \frac{1}{2}\rho V^2 S K_G}, \quad (6.10)$$

where $f_0 = F_0/W$, and if $(C_L)_R$ is the lift coefficient for steady level flight at the speed V_R, an explicit form of Equation (6.10) is found to be

$$s_R = -[w/(g\rho K_G)] \ln[1 - K_G/\{(C_L)_R(f_0 - \mu_R)\}]. \quad (6.11)$$

This may also be expressed as a series:

$$s_R = \frac{w}{g\rho(C_L)_R(f_0 - \mu_R)} \left[1 + \frac{1}{2} \frac{K_G}{(C_L)_R(f_0 - \mu_R)} \right.$$
$$\left. + \frac{1}{3} \frac{K_G^2}{(C_L)_R^2(f_0 - \mu_R)^2} + \cdots \right], \quad (6.12)$$

where the factor outside the brackets in this equation is the expression obtained from Equation (6.4) with F replaced by $(F_0 - \mu_R W)$, i.e. assuming a constant thrust F_0 and making a correct allowance for rolling resistance but with no allowance for aerodynamic lift or drag. The series within the brackets in Equation (6.12) represents the effects of lift, drag and variation of thrust and the correction to s_R given by the sum of the series is shown in Figure 6.9, for an aircraft with $C_{DG} = 0.05$, $C_{LG} = 0.4$, $\mu_R = 0.025$, $w = 5 \text{ kN/m}^2$ and $(C_L)_R = 1.25$, corresponding to a rotation speed V_R of 81 m/s. From the curves given in Figure 5.7 it can be shown that the values of k_1 giving the best agreement with the data up to a speed of 80 m/s are in the region of 4×10^{-5} (m/s)$^{-2}$ for a typical civil turbofan and 6×10^{-5} (m/s)$^{-2}$ for a propfan. Except when k_1 is much smaller than these values, as it may be for a military turbofan, the main contribution to the correction shown in Figure 6.9 comes from the reduction of thrust with increasing speed. In all cases the correction becomes nearly independent of f_0 when f_0 is large.

It has been explained in § 5.2.2, 5.2.4 and 5.5.1 that for civil and military turbofans and for propfans a better representation of the loss of thrust with increasing forward speed is given by Equation (5.8). If this equation is used to substitute for F in Equation (6.5) an equation for s_R is obtained which is more accurate than Equation (6.8), but the integration is not as simple because the net accelerating force is no longer a simple linear function of V^2. When Equation (5.8) is used, the best method of estimating the distance s_R is by means of a step-by-step computation and it is convenient to divide the whole range of velocity, from zero to V_R, into equal velocity steps ΔV, the distance segments over which each step occurs being initially quite short but becoming longer as V_R is approached. Then, if a linear increase of velocity is assumed within each segment the velocities $\bar{V}$ at the mid-points are $\Delta V/2$, $3\Delta V/2$, $5\Delta V/2$... and in each segment the mean acceleration is $\bar{V}\Delta V/\Delta s$, where Δs is the distance covered in the segment. The net accelerating force at the mid-point of the nth segment is

$$X_{Gn} = F - D - \mu_R(W - L)$$

and since

$$\Delta s = m\bar{V}\,\Delta V / X_G$$

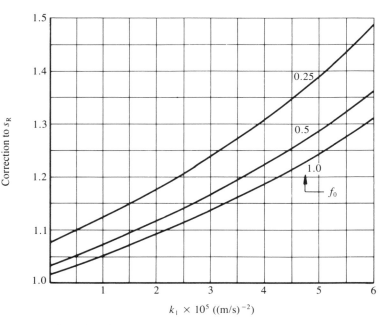

Figure 6.9. Correction to s_R given by series in Equation (6.12). $C_{DG} = 0.05$, $C_{LG} = 0.4$, $\mu_R = 0.025$, $(C_L)_R = 1.25$, $w = 5\,\text{kN/m}^2$.

The take-off ground run

the total distance is

$$s_R = \tfrac{1}{2}m(\Delta V)^2[1/X_{G1} + 3/X_{G2} + 5/X_{G3} \\ + \ldots + (2n-1)/X_{Gn} + \ldots]. \tag{6.13}$$

Obviously the accuracy increases with the number of segments but it is found that a number as low as 10 gives quite acceptable accuracy in s_R.

During the ground run with $V \leq V_R$ the angle of incidence is determined by the design of the undercarriage and by the distribution of load on it, the latter depending to some extent on the position of the CG. There will be other constraints on the design of the undercarriage but nevertheless an attempt can be made to design for maximum accelerating force X_G. Equation (6.5) shows that the required condition is that $(C_D - \mu_R C_L)$ should be a minimum and if $C_D = K_1 + K_2 C_L^2$ the incidence should be chosen to make

$$K_2 C_L^2 - \mu_R C_L \text{ minimum.}$$

Thus the optimum value of C_L during the ground run is $\mu_R/(2K_2)$. The value of μ_R that is usually assumed for a concrete runway is 0.025 and with a substantial favourable ground effect the value of K_2 might be about 0.03 in a typical case, giving an optimum C_L of about 0.4. In predicting the required field length it may be inadvisable to assume that this optimum value of C_L is always achieved, because it cannot be maintained at all the possible positions of the CG. To ensure that the predicted distance s_R is not too optimistic the fuselage attitude should be determined for a range of CG positions, perhaps allowing also for variation of pitching moment due to engine thrust.

When an engine failure occurs during the ground run there is a sudden change of thrust and the distance s_R is most easily estimated in two stages, one before and one after the engine failure. The length required to accelerate from rest to the engine failure speed V_{EF} can be found approximately from Equation (6.6), replacing V_R by V_{EF} and calculating $\bar{a}$ from Equation (6.5) by putting $V = 0.7 V_{EF}$. A more accurate estimate can be obtained by integration and if Equation (6.7) is acceptable the equation to be used is (6.11) with $(C_L)_R$ replaced by $(C_L)_{EF}$, the lift coefficient that would be required for steady level flight at the speed V_{EF}. (The lift coefficient $(C_L)_{EF}$ is being used here only as a convenient substitution for $2w/(\rho V_{EF}^2)$ and the procedure is valid even if V_{EF} is so low that $(C_L)_{EF}$ could not be reached because it would be greater than C_{Lmax}.) If Equation (5.8) is used instead of Equation (6.7) the distance required to accelerate to the speed V_{EF} can be found most easily by the step-by-step computation already explained.

The length required in the second stage, to accelerate from the speed V_{EF} to V_R, can be found by integration of Equation (6.5) between the appropriate limits, using a thrust equation such as

Equation (6.7) or (5.8) and allowing for the reduction of thrust caused by engine failure. If Equation (6.7) is used the length is given by Equation (6.10) with the limits changed and this leads to a modified form of Equation (6.11), viz.

$$s = -\frac{w}{g\rho K_G} \ln\left[\frac{1 - K_G/[(C_L)_{EF}(f_0 - \mu_R)]}{1 - K_G/[(C_L)_R(f_0 - \mu_R)]}\right]. \quad (6.14)$$

An alternative method for use when Equation (6.7) is valid is described in ESDU EG 5/1 (1972). A chart is given showing the speed at which a mean acceleration $\bar{a}$ should be calculated from Equation (6.5). The distance required to accelerate from the speed V_{EF} to V_R is then obtained from that mean acceleration as

$$s = \tfrac{1}{2}(V_R^2 - V_{EF}^2)/\bar{a}. \quad (6.15)$$

If a step-by-step computation is to be made for this part of the ground run, using any form of thrust relation such as Equation (5.8), the range of velocity from V_{EF} to V_R is divided into equal velocity steps ΔV as explained earlier and it can be shown that the total distance is

$$s = mV_{EF}\Delta V[1/X_{G1} + 1/X_{G2} + 1/X_{G3} + \ldots]$$
$$+ \tfrac{1}{2}m(\Delta V)^2[1/X_{G1} + 3/X_{G2} + 5/X_{G3} + \ldots], \quad (6.16)$$

where ΔV and $X_{G1}, X_{G2}\ldots$ have the same meanings as in Equation (6.13). The form of Equation (6.16) is different from that of Equation (6.13) because the computation now starts from $V = V_{EF}$ and not from $V = 0$.

After the distance s_R has been found it is still necessary to estimate the distance s_{RL} covered during the rotation phase, in order to find the total length of ground run s_g. In principle s_{RL} could be estimated from a knowledge of the speeds V_R and V_{LO} and the mean acceleration during the rotation phase, but there are some difficulties in this approach. Both C_L and C_D increase strongly in this phase of the take-off and any estimates of their mean values would be uncertain. Moreover, although the speed V_R is directly under the control of the pilot and can be specified exactly, the lift-off speed V_{LO} depends on V_R and on the rate of rotation and may not be known in advance. Recognising these difficulties Williams (1972) suggests that the distance s_{RL} should be found by assuming a value for the time t_R taken for the rotation phase. This time can be estimated approximately from experience with other comparable aircraft and is about 3 seconds for a typical subsonic transport. When a value has been decided for t_R a rough estimate of the mean acceleration in the rotation phase can be used to find V_{LO}. Since this speed is usually only a little greater than V_R the mean speed can be found with acceptable accuracy even if the

The take-off ground run

error in the estimated acceleration is quite large. With the mean speed known the distance s_{RL} can be found from t_R and if s_R has been calculated the total length of ground run s_g can be found.

EXAMPLE 6.1. TAKE-OFF GROUND RUN

An aircraft takes off with a mass of 60 000 kg in the ISA at sea level. The wing area is 135 m², $\mu_R = 0.025$ and the rotation speed V_R is chosen to be 75 m/s. (This value of V_R is a reasonable choice if $C_{Lmax} = 1.55$, since V_R is required to be at least 1.1 V_{MU} and $V_{MU} \approx V_s$. For $C_{Lmax} = 1.55$ the minimum value of V_R is therefore about $1.1[2W/(\rho S C_{Lmax})]^{1/2} = 74.5$ m/s.) During the ground run, for $V \leq V_R$, $C_L = 0.4$ and $C_D = 0.05$. The turbofan engines have a total static thrust F_0 of 150 kN and the variation of thrust with speed is given with good accuracy by Equation (5.8) with $k_2 = 2.5 \times 10^{-3}$ (m/s)$^{-1}$ and $k_3 = 4.5 \times 10^{-6}$ (m/s)$^{-2}$.

(1) Find s_R from Equation (6.13), using 10 steps.
(2) Choose a value of k_1 in Equation (6.7) to give agreement with Equation (5.8) at $V = 0.8 V_R$. Then find s_R from Equation (6.11).
(3) Find s_R from the mean acceleration, calculated for $V = 0.7 V_R$, using Equation (5.8) to give the thrust.
(4) Find the thrust work done in accelerating from rest to the speed V_R, using the mean thrust from (3) above. Hence express the kinetic energy at $V = V_R$ as a percentage of the thrust work.
(5) For the rotation phase, between the speeds V_R and V_{LO}, assume $t_R = 3$ seconds and mean coefficients $C_L = 0.9$ and $C_D = 0.08$. Find the mean acceleration, the lift-off speed V_{LO} and the distance travelled during rotation. Hence estimate the length of the total take-off ground run s_g.

Preliminary calculations give the following values: $W = 5.886 \times 10^5$ N, $w = W/S = 4360$ N/m², $\frac{1}{2}\rho S = 82.69$ kg/m, $C_{DG} - \mu_R C_{LG} = 0.040$, $f_0 = 0.2548$ and $f_0 - \mu_R = 0.2298$.

(1) $X_G = F - D - \mu_R(W - L)$
$= F_0(1 - k_2 V + k_3 V^2) - \frac{1}{2}\rho V^2 (C_{DG} - \mu_R C_{LG}) - \mu_R W$
$= (a - bV - cV^2) \times 10^3$,

where $a = 135.3$, $b = 0.375$ and $c = 0.00263$. Using Equation (6.13) with 10 steps,

$$\Delta V = 7.5 \text{ m/s and for the } n\text{th step } \bar{V} = 3.75(2n - 1).$$

Thus

$$s_R = 30 \times (7.5)^2 \sum_{1}^{10} \left(\frac{2n-1}{a - bV - cV^2}\right) = 1560 \text{ m}.$$

(If 20 steps are used the result differs by only 1 m.)

(2) Equating the two thrust functions,
$$1 - k_1 V^2 = 1 - k_2 V + k_3 V^2$$
and if this is to be satisfied when $V = 0.8 V_R = 60$ m/s,
$$k_1 = k_2/V - k_3 = 3.72 \times 10^{-5} \text{ (m/s)}^{-2}.$$
Equation (6.9) then gives $K_G = 0.1074$ and the lift coefficient required for level flight at the speed V_R is $(C_L)_R = 2w/(\rho V_R^2) = 1.265$. Finally, Equation (6.11) gives $s_R = 1558$ m.

(3) For $V = 52.5$ m/s, Equation (5.8) gives $F/F_0 = 0.8812$. Equation (6.5) then gives the mean acceleration as
$$\bar{a} = g[0.8812 f_0 - 0.025 - \tfrac{1}{2}\rho V^2 S(C_D - \mu_R C_L)/W]$$
$$= 1.8054 \text{ m/s}^2$$
and hence
$$s_R = \tfrac{1}{2} V_R^2/\bar{a} = 1558 \text{ m}.$$

The results (1), (2) and (3) agree within 0.2%, but it should be noted that the agreement of (2) with (1) and (3) would be less satisfactory if the Equations (6.7) and (5.8) were matched at a speed less than $0.8 V_R$. In general it is wiser to determine the constants in separate curve-fitting exercises than to assume an equivalence of the two functions at one point. The procedure used here has been adopted for the sake of simplicity, but of course any curve-fit should be aimed at accuracy in the important region and it is reasonable to aim for a good fit in a speed range somewhat below V_R.

(4) Thrust work $= 0.8812 \times 150 \times 10^3 \times 1560 = 2.06 \times 10^8$ J. When $V = V_R$, kinetic energy $= \tfrac{1}{2} m V_R^2 = 1.69 \times 10^8$ J. This is 81.9% of the thrust work, showing that the work done in overcoming rolling resistance and even aerodynamic drag is a small proportion of the total thrust work.

(5) For the rotation phase the calculation will be made for $V = V_R$ as a first approximation. Equation (5.8) gives $F = 0.838 F_0$ and Equation (6.5) gives the mean acceleration as
$$\bar{a} = g[0.838 f_0 - 0.025 - \tfrac{1}{2}\rho V^2 S(C_D - \mu_R C_L)/W]$$
$$= 1.403 \text{ m/s}^2.$$
Hence $V_{LO} = 75 + (3 \times 1.403) = 79.2$ m/s.

In order to obtain a more accurate mean velocity a second approximation is made using $V = 77.1$ m/s and this gives $\bar{a} = 1.368$ m/s^2, which leads to a lift-off speed only slightly reduced to 79.1 m/s. The distance travelled during rotation is then found as $3 \times \tfrac{1}{2}(V_R + V_{LO}) = 231$ m and the total length of the ground run is $s_g = 1560 + 231 = 1791$ m.

6.8 Lift-off, transition and climb

Referring to Figure 6.6, there is a transition phase after lift-off during which the flight path is curved upward and the lift is greater than the weight. After the transition the aircraft climbs steadily at an angle γ_c which may be assumed to remain constant, at least for a few seconds. The point at which the screen height h_{sc} is reached may be either before or after the end of the curved transition and the horizontal distance to this point from the point of lift-off is the distance s_a which is required for the determination of the total take-off distance $(s_g + s_a)$.

During the transition there is some increase of speed and there is also a variation of lift coefficient, so that the drag does not remain constant. Nevertheless it is often assumed, for the purpose of estimating the distance s_a, that the motion of the aircraft can be represented adequately by specifying mean values of the speed and lift coefficient and regarding these as constants. Then if β is the ratio D/L in the steady climb after transition, the angle of this climb is

$$\gamma_c = f - \beta,$$

if the angle is assumed to be small and if any acceleration during the climb is neglected.

A very rough approximation to the distance s_a may be obtained by neglecting completely the horizontal distance covered in the curved transition and assuming that the steady climb at the angle γ_c is established immediately after lift-off. This gives

$$s_a = h_{sc}/\gamma_c,$$

but the true value of s_a is usually much greater than this because the distance covered in the transition before achieving the climb angle γ_c is a substantial proportion of s_a.

A more realistic assumption is that the flight path during the transition is a circular arc of radius R. Referring to Figure 6.10(a), the height reached by the circular arc when the climb angle reaches γ_c is

$$R(1 - \cos \gamma_c) = \tfrac{1}{2} R \gamma_c^2, \text{ if } \gamma_c \text{ is small,}$$

so that the condition for the screen height h_{sc} to be reached before the end of the transition to a climb angle γ_c is

$$\gamma_c > (2 h_{sc}/R)^{1/2}.$$

Reference to Figure 6.10(b) shows that in this case the angle of climb γ at the screen height is related to h_{sc} by

$$h_{sc} = R(1 - \cos \gamma)$$

and the horizontal length of the airborne segment is

$$s_a = R \sin \gamma.$$

From these equations it is easily shown that in this case, if γ is small,

$$s_a = (2Rh_{sc})^{1/2}. \tag{6.17}$$

For the case shown in Figure 6.10(a), where the screen height is not reached until after the end of the curved transition, the distance OA is $\frac{1}{2}\gamma_c R$ if γ_c is small and hence

$$s_a = \tfrac{1}{2}\gamma_c R + h_{sc}/\gamma_c. \tag{6.18}$$

With γ_c found from known values of f and β, the distance s_a can be calculated from Equation (6.17) or (6.18) provided that the radius R can be estimated. The radius can be expressed in terms of the mean speed V and the ratio $L/W = n$ by writing

$$mV^2/R = L - W = W(n-1)$$

and thus

$$R = V^2/[g(n-1)]. \tag{6.19}$$

Alternatively the radius can be expressed in terms of the lift coefficient of the aircraft during the transition. If this is written as $(C_L + \Delta C_L)$, where C_L is the lift coefficient for steady level flight at the mean speed V,

$$\Delta C_L = (n-1)w/(\tfrac{1}{2}\rho V^2)$$

and thus

$$R = 2w/(\rho g\, \Delta C_L). \tag{6.20}$$

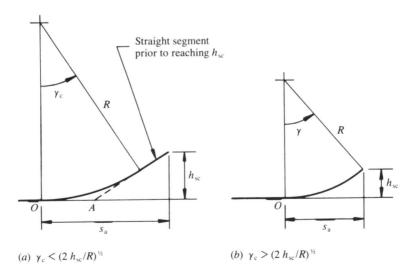

(a) $\gamma_c < (2h_{sc}/R)^{1/2}$ (b) $\gamma_c > (2h_{sc}/R)^{1/2}$

Figure 6.10. Airborne part of take-off path.

The equations (6.19) and (6.20) show that the assumptions of constant speed and constant radius imply constant values of n and ΔC_L. For the calculation of the distance s_a from the equations given here it is necessary to assume a value for either n or ΔC_L, but unfortunately this is difficult because the values of n and ΔC_L that are achieved depend on the piloting technique. Buckingham & Lean (1954) have analysed measurements made in take-off tests of a number of aircraft. In these tests the take-offs were of the 'maximum effort' type, with the pilot trying to keep the increase of speed after lift-off as small as possible, since this technique would be expected to give the shortest possible airborne distance. As might be expected, Buckingham & Lean found that the value of ΔC_L employed by the pilot increased with the maximum lift coefficient available and also with the ratio of the mean speed in the transition to the stalling speed. The pilot would recognise that if his transition speed was well above the stalling speed the lift coefficient would be well below C_{Lmax} and thus the safe additional incidence available would be substantial. For these 'maximum effort' take-offs Buckingham & Lean found that the increment of lift coefficient could be represented approximately by the empirical equation

$$\Delta C_L = [(V/V_s)^2 - 1][C_{Lmax}\{(V_s/V)^2 - 0.53\} + 0.38]. \tag{6.21}$$

Figure 6.11 shows ΔC_L as given by this equation, plotted against V/V_s for three values of C_{Lmax}. Buckingham & Lean suggest that in a normal take-off the ΔC_L employed by a pilot is likely to be about half the value obtained in a maximum effort take-off, so that the values of ΔC_L given by Equation (6.21) or Figure 6.11 should be divided by 2 for normal take-offs.

Equations (6.17) and (6.20) show that when $\gamma_c > (2h_{sc}/R)^{1/2}$, so that the screen height h_{sc} is reached before the end of the transition,

$$s_a = [4wh_{sc}/(\rho g \, \Delta C_L)]^{1/2}. \tag{6.22}$$

If $\gamma_c < (2h_{sc}/R)^{1/2}$, so that the screen height is reached after the end of the transition, Equations (6.18) and (6.20) show that

$$s_a = \gamma_c w/(\rho g \, \Delta C_L) + h_{sc}/\gamma_c. \tag{6.23}$$

As an example, Figure 6.12 shows the distance s_a as a function of ΔC_L and wing loading for a screen height of 10.67 m (35 feet) and $\gamma_c = 0.10$. The full lines refer to the case where the screen height is reached before the end of the transition and are valid for all values of γ_c that are not less than 0.10. The broken lines refer to the case shown in Figure 6.10(a) and would be different for other values of γ_c.

Other methods have been used for calculating the distance s_a, based on different forms of theoretical model. Like the method already

144   *Take-off and landing performance*

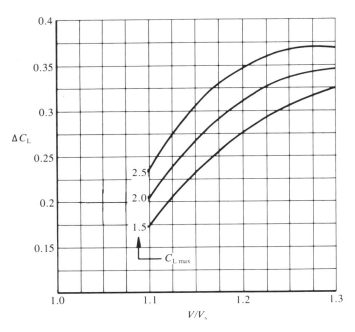

Figure 6.11. Increment of C_L in 'maximum effort' take-off transition, from Equation (6.21).

discussed these all rely on experimental data and may therefore be regarded as convenient frameworks for analysis of the measured data. In one method it is assumed that the lift coefficient remains constant at the value for steady level flight at the lift-off speed V_{LO}, so that the excess lift required to produce curvature of the flight path is all caused by the increase of speed above V_{LO}. This approach was first described in 1940 in an unpublished paper by Ewans and Hufton but it has been explained and discussed more fully by Perry (1969a). In a further paper Perry (1969b) has given a method of estimating s_a based on the more realistic assumption that the aircraft is rotated at a constant angular velocity of pitch from the moment of lift-off to the point at which the steady climb angle γ_c is attained.

Reference has been made in § 6.6 to the division of the airborne part of the take-off and the subsequent climb into segments for the purpose of airworthiness certification. A specified climb gradient is required in each segment after failure of one engine and the requirements have been summarised by Williams (1972).

EXAMPLE 6.2. AIRBORNE PART OF TAKE-OFF DISTANCE

For the aircraft considered in Example 6.1 the stalling speed V_s is 68 m/s and in a normal take-off with all the engines operating

Lift-off, transition and climb

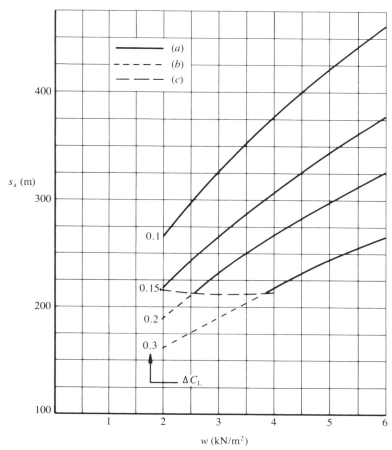

Figure 6.12. Airborne distance s_a in take-off, $\gamma_c = 0.10$, $h_{sc} = 10.67$ m. (a) Screen height reached before end of transition. (b) Screen height reached after end of transition. (c) Condition for screen height to be reached exactly at end of transition.

$V_{LO} = 79$ m/s and the speed V_3 at the screen height of 10.67 m is 87 m/s. (This value of V_3 must be significantly greater than $1.2 V_s$ ($= 81.6$ m/s) in order to ensure that even after an engine failure the speed at the screen height is not less than $1.2 V_s$.) If the mean drag coefficient in the transition is estimated to be 0.125 and ΔC_L is assumed to be half the value given by Equation (6.21), estimate the horizontal distance s_a from lift-off to the point where the screen height is reached.

The aircraft has four engines and one of them fails at lift-off, so that one quarter of the thrust is lost instantaneously. Assuming that in this case the speed at the screen height is only 82 m/s and the drag/lift ratio β is unchanged, find the new value of s_a.

The mean speed during the transition is

$$V = \tfrac{1}{2}(79 + 87) = 83 \text{ m/s}$$

and the lift coefficient in steady flight at this speed is

$$C_L = 2w/(\rho V^2) = 1.033.$$

Then, calculating the quantities required for use in Equation (6.21),

$$V/V_s = 1.221, \quad C_{Lmax} = 2w/(\rho V_s^2) = 1.539$$

and the increment of C_L is found from the equation as

$$\Delta C_L = 0.293,$$

but the value to be used is only half this value, i.e.

$$\Delta C_L = 0.1465.$$

At $V = 83$ m/s Equation (5.8) gives $F/F_0 = 0.8235$ and hence

$$f = F/W = 0.210.$$

In a steady climb with $C_L = 1.033$, $\beta = C_D/C_L = 0.125/1.033 = 0.121$ and hence the attainable angle of steady climb is

$$\gamma_c = f - \beta = 0.089.$$

Equation (6.20) gives $R = 2w/(\rho g \, \Delta C_L) = 4953$ m.

The criterion for comparing the end of the curved transition with achievement of the screen height is

$$(2h_{sc}/R)^{1/2} = 0.0656$$

and since this is less than γ_c the screen height h_{sc} is reached before the end of the transition. Equation (6.17) is therefore applicable and

$$s_a = (2Rh_{sc})^{1/2} = 325 \text{ m}.$$

When one engine fails at lift-off and the speed at the screen height is only 82 m/s the mean speed during the transition is

$$V = \tfrac{1}{2}(79 + 82) = 80.5 \text{ m/s}$$

and the lift coefficient for steady flight at this speed is 1.098. Then, since $V/V_s = 1.184$ the increment of C_L to be used is found to be 0.133, using half the value given by Equation (6.21). At the speed of 80.5 m/s Equation (5.8) gives $F/F_0 = 0.828$ and after failure of one engine $F_0 = 3/4 \times 150 = 112.5$ kN, giving $f = F/W = 0.1583$. The drag/lift ratio β is assumed to be unchanged at 0.121 and the attainable angle of steady climb is therefore

$$\gamma_c = f - \beta = 0.0373.$$

The new transition curve radius is $R = 5456$ m and the criterion $(2h_{sc}/R)^{1/2}$ becomes 0.0625. This is greater than γ_c, showing that in this case the screen height h_{sc} is not reached until after the end of the transition and therefore Equation (6.18) is applicable, giving

$$s_a = \tfrac{1}{2}\gamma_c R + h_{sc}/\gamma_c = 388 \text{ m}.$$

The landing approach

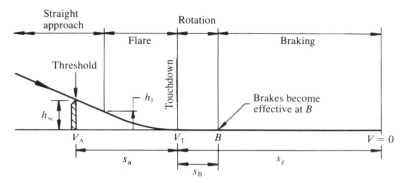

Figure 6.13. Landing trajectory segments (not to scale).

6.9 Landing procedure

Figure 6.13 illustrates the trajectory used for landing. The aircraft approaches the runway initially along a straight path, but if this straight approach were continued all the way to the ground the vertical component of velocity on impact would be unacceptably high and it is necessary to introduce a 'flare' in which the flight path curves upward. In the flare the pilot aims to reduce the vertical component of velocity so that it is no more than about 0.5 m/s when the main wheels touch the runway, although the undercarriage is usually designed for a much higher vertical velocity component, of order 3 m/s. The point at which this occurs is known as the touchdown point, but at this stage the nose wheel is still well above the runway and it is necessary to rotate the aircraft in the nose-down sense before effective braking can be applied. In the final phase of the landing the aircraft is brought to rest by use of the wheel brakes, sometimes assisted by lift dumpers or spoilers to reduce the wing lift and thus increase the ground reaction, while also increasing the drag, and reversed thrust from the engines even on some propeller-driven aircraft. On military aircraft braking parachutes are sometimes used.

6.10 The landing approach

The undercarriage is lowered and the flaps are set to the landing configuration while the aircraft is at least 300 m above the level of the runway. The aircraft is then held in a straight glide path, usually at an angle of 3° to the horizontal, and as the height falls below about 60 m the speed is gradually reduced. At a specific threshold or screen height, usually taken to be either 9.1 or 15.2 m, the approach speed is required to be not less than 1.3 times the stalling speed V_s. The flare is started at the *flare point* which is usually after the threshold with a 3° approach but may be before it with steeper approaches.

6.11 The landing flare

As mentioned earlier, the pilot's aim in the flare is to reduce the vertical component of velocity to a very low value at the touchdown point. As this point is approached the increasing effects of ground proximity lead to a reduction of C_D and an increase of C_L for a given incidence, so that the glide path tends to become less steep. For this reason it is found with some aircraft that the required flare is induced automatically by the effect of the ground, so that little or no action by the pilot is needed. It may also happen, if the flare manoeuvre is misjudged or started at too great a height, that the vertical component of velocity falls to zero while the aircraft is still at a small height above the ground. There is then a period of 'float', in which the speed is allowed to fall and the aircraft gradually descends on to the ground. Provided there is a sufficient length of runway available, a landing of this kind is safe and much to be preferred to a flare which is misjudged in the opposite sense, so that the aircraft hits the ground with a substantial vertical component of velocity. Since any float at the end of the flare increases the total landing distance it will be assumed in the following analysis that the flare is perfectly judged so that there is no float and the touchdown occurs with zero vertical component of velocity.

It is to be expected on simple theoretical grounds that the landing distance required will always be much less than the take-off distance, because braking and reversed thrust can impart to a relatively light aircraft a deceleration that is greater than the acceleration imparted by engines alone to an aircraft with a heavy fuel load. It is found in practice, however, that the runway length required for landing can be much greater than the length given by simple calculations, because of the problem of controlling the flare trajectory and the possibility of float before touchdown. These cause considerable variation in the position of touchdown and hence increase the length of runway that must be provided.

Figure 6.14 is based on a report by Pinsker (1975) of computer studies by Lockheed on the spread of touchdown positions in landings of the L1011 (Tristar). The upper diagram shows the spread occurring with the normal Autoland system, an automatic pilot for use in landing that gives results closely similar to those obtained with manual operation of the controls. In this case only 19 out of 20 landings achieve a touchdown that is even within a spread of 440 m and the spread shown at (c) indicates that the length of runway that may be actually required is about 700 m greater than the length given by simple calculations, because of the uncertainty of the touchdown position. The lower diagram in Figure 6.14 shows that the spread of touchdown positions can be greatly reduced by incorporating Direct

The landing flare

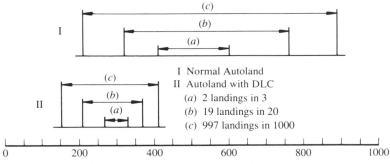

Figure 6.14. Spread of touchdown positions for Lockheed 1011, from Pinsker (1975).

Lift Control (DLC) in the Autoland system. With DLC the control surfaces of the aircraft are operated in an unconventional manner to obtain rapid changes of C_L on demand, without rotation of the whole aircraft. This makes it possible to control the trajectory in the flare much more accurately, so reducing the spread of touchdown positions. Without the aid of DLC such precise control of the flare trajectory is not possible and some landings will be made with a substantial float distance, whereas in others there will be an appreciable vertical component of velocity at touchdown.

ESDU EG 6/3 (1960) describes a method that has been commonly used for estimating the airborne landing distance s_a shown in Figure 6.13. It is assumed that the flight path in the flare is a circular arc whose radius can be calculated with sufficient accuracy by considering flight at a constant speed V_F which is equal to the mean speed in the flare. It is also assumed that the speed at the point where the flare is started is the same as the approach speed V_A at the threshold, so that if V_T is the speed at touchdown the mean speed in the flare is

$$V_F = \tfrac{1}{2}(V_A + V_T).$$

The analysis is essentially the same as that used for calculating the length of the airborne path in a take-off, assuming a circular arc transition, and as in that case the assumed value of n or ΔC_L is of great importance. ESDU EG 6/3 (1960) recommends that n should be taken as 1.1 and that for a first approximation V_T/V_A should be assumed to be 0.9. With this value of n, Equation (6.19) gives the radius of curvature as

$$R = V^2/(0.1\,g) \qquad (6.24)$$

and if γ_A is the approach glide angle in radians the horizontal distance covered in the flare is

$$s_F = R\gamma_A.$$

The height at the flare point, where the flare is started, is

$$h_f = R(1 - \cos \gamma_A) = \tfrac{1}{2}R\gamma_A^2$$

and in the usual case where $h_f < h_{sc}$ the total airborne distance from the threshold to touchdown is

$$s_a = (h_{sc} - h_f)/\gamma_A + R\gamma_A = h_{sc}/\gamma_A + \tfrac{1}{2}R\gamma_A \tag{6.25}$$

and this is the same as Equation (6.18), with γ_c replaced by γ_A.

The exceptional case where $h_f > h_{sc}$ may arise occasionally when steep approach angles are used. Figure 6.10(b) is then relevant and Equation (6.17) gives the distance s_a from the threshold to touchdown.

ESDU EG 6/3 (1960) explains how the first approximation for V_T/V_A may be improved by iteration, provided that a reasonable estimate of the mean net drag $(D - F)$ can be made. The work done against the net drag over the distance s_a can be equated to the loss of potential and kinetic energy, giving

$$(D - F)s_a = h_{sc}W + (W/2g)(V_A^2 - V_T^2)$$

and hence

$$V_A^2 - V_T^2 = 2g[(\beta - f)s_a - h_{sc}]. \tag{6.26}$$

Thus a new value of V_T can be obtained and s_a can be recalculated. Further iterations can be made if required but are usually not justified because of the inaccuracy of the estimated mean net drag. It is important, however, that Equation (6.26) should be used to give the best possible estimate of V_T for use in calculating the length of ground run.

An alternative analytical model has been proposed by White (1968) for the landing of transport aircraft. This allows for some inaccuracy in the pilot's judgement of the flare by specifying three phases, (i) an initial flare, (ii) a float period and (iii) a touchdown period. Pinsker (1969) has studied the flare in some detail, concentrating on the actions taken by typical pilots and discussing the frequent occurrence of a float period. Seckel (1975) has made theoretical studies and has also examined a large number of experimental landings with a variable stability aircraft, with the aim of determining the characteristics required to make a light aircraft easy to land safely. Some other methods of calculating the airborne landing distance are related to landing techniques which are specified precisely by airworthiness certification authorities, e.g. the Reference Landing Distance specified by the Joint Airworthiness Authority in JAR (BB)-25.125.

6.12 The landing ground run

Figure 6.13 shows a period of rotation after touchdown which introduces a slight delay before the wheel brakes can become effective. The typical rate of rotation is about 3° per second and the time delay is usually about 2–3 seconds. Data given in ESDU 85029 (1985) show that there is some deceleration during the delay period, even though the brakes have not yet become effective, but in a typical case the speed falls by only about 1.5% per second. Before proceeding to calculate the length of the ground run, after the point B where the brakes become effective, estimated values of the delay time and deceleration should be used to calculate the speed V_B at B and the distance s_B from touchdown to B.

The equation of motion to be used for calculating the length of the ground run after the point B is essentially the same as the Equation (6.5) used for take-off, but it will be written in a different form here because the maximum available value of the braking force G varies with speed and cannot be expressed simply as $\mu_B(W - L)$, where μ_B is a constant coefficient of friction. With X_G used as in the take-off analysis to denote the net accelerating force (negative in this case) the equation for a level runway corresponding to the earlier Equation (6.5) is

$$m \, dV/dt = mV \, dV/ds = X_G = F - D - G. \tag{6.27}$$

When thrust reversers are not used the engines are run at idling speed throughout the ground run and the thrust F is usually small enough to be neglected. In cases where reversed thrust can be used, it is generally applied after the spoilers and brakes become effective and thus after nose-wheel steering can be effective. The negative value of F then makes a significant contribution to the overall negative value of X_G. As the speed falls below a certain specified limiting value, usually between about 20 and 40 m/s, it is necessary to cancel the thrust reversal in order to avoid the possibility of reingestion of hot exhaust gas into the engine intake.

Except in some cases where a braking parachute is used the drag force D in Equation (6.27) is determined by a nearly constant coefficient C_D throughout the ground run. The coefficient is considerably higher than it is in the take-off ground run because of the greater flap angle that is used for landing, as shown in Figure 6.3. As already mentioned, spoilers or lift dumpers can be used to give a useful increase of C_D and reduction of C_L. Braking parachutes are often used on military aircraft and are deployed immediately after the point B in Figure 6.13. The parachute is sometimes deployed in a 'reefed' condition at first, to reduce the initial shock load, the canopy being fully opened only after there has been some reduction of speed.

With this arrangement there is of course a variation of C_D during the ground run. The use of braking parachutes is discussed in ESDU 85029 (1985).

The most important term contributing to X_G is the braking force G, defined as the component of the force exerted by the ground on the wheels, in the plane of the runway surface and parallel to the direction of motion. The principles on which this force depends are explained in ESDU 71025 (1988) and extensive data are given in ESDU 71026 (1988). If a coefficient of braking friction is defined as $\mu_B = G/Z$, where G is the braking force on one wheel and Z is the load on that wheel normal to the runway, it is found that the maximum possible value of μ_B occurs when the tyre is on the point of skidding. Figure 6.15 is based on data for typical concrete runways given in ESDU 71026 (1988) and shows how the average maximum value of μ_B decreases with increasing aircraft speed V. When the runway is dry the decrease is linear and fairly small but with a wet surface there is a

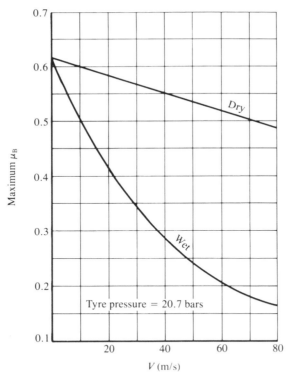

Figure 6.15. Average braking coefficient of friction between tyre and concrete runway, from ESDU 71026 (1988).

The landing ground run

large decrease and the variation is not linear. The curve shown for the wet runway refers to a tyre with a ribbed tread; with a smooth tread the decrease of maximum μ_B with increasing V is even more severe, whereas on a dry runway the form of the tread makes no difference. Because skidding cannot be allowed to occur, the value of μ_B that can be achieved in practice, even with the best form of automatic braking system, is usually no more than about 80% of the maximum possible value. Since braking is usually applied only to the main wheels and not to the nose wheel, the estimation of the force G is further complicated by the effect of strong deceleration in transferring the ground reaction from the main wheels to the nose wheel and thus reducing the braking capacity.

Apart from the limitation imposed by the maximum value of μ_B, there is another important factor that may limit the value of G. This is the *brake torque limit*, the maximum torque that can be exerted on a wheel by the braking system. This usually limits the maximum value of G that can be obtained to about 0.35 W, a value that could be exceeded on a dry runway if incipient skidding were the only limiting factor, especially at low speeds where Figure 6.15 shows $\mu_B \approx 0.6$. Even with only 80% of this made available in practice the value of G could approach 0.5 W. It should also be noted that in many landings neither the skidding limit nor the torque limit is approached by normal pilot action, simply because there is normally no urgency for a rapid stop and if reversed thrust is used as the dominant decelerating force the brake wear can be reduced considerably. There are many documented cases involving runway incidents where pilot action did not approach the allowable limits until real danger became evident. Nevertheless, for proof of potential performance it is necessary to make certain assumptions about braking capacity and how closely this approaches the maximum available.

The simplest approximate estimate of the distance $s_g - s_B$ is obtained by estimating the mean acceleration $\bar{a}$ and using this in an equation similar to Equation (6.6), viz

$$s_g - s_B = -\tfrac{1}{2} V_B^2 / \bar{a}. \tag{6.28}$$

ESDU EG 6/4 (1971) suggests that if this method is used the mean acceleration should be calculated for $V = 0.7 V_T$, but if an allowance is made, as already discussed, for the delay period this should become $V = 0.7 V_B$.

For a more exact solution, the required distance can be found from Equation (6.27) by integration. The equation corresponding to Equation (6.8) is

$$s_g - s_B = -\tfrac{1}{2} m \int_0^{V_B} \frac{d(V^2)}{X_G}. \tag{6.29}$$

In most cases there is no way in which X_G can be represented satisfactorily as an analytic function of V^2 and it is necessary to evaluate the integral by a step-by-step computation. As in the method described earlier for calculating the take-off ground run, it is convenient to divide the range of velocity from V_B to 0 into equal steps ΔV. Then for successive segments the velocities $\bar{V}$ at the mid-points are $(V_B - \Delta V/2)$, $(V_B - 3\Delta V/2)$, $(V_B - 5\Delta V/2)$, ... and since the distance covered in the nth step is

$$\Delta s = -m\bar{V}_n \Delta V / X_{Gn}$$

the total distance is

$$s_g - s_B = -m\,\Delta V[(V_B - \Delta V/2)/X_{G1} + (V_B - 3\Delta V/2)/X_{G2}$$
$$+ \ldots + (V_B - (2n-1)\Delta V/2)/X_{Gn} + \ldots]. \quad (6.30)$$

If the total number of steps is r it can be shown that Equation (6.30) can also be written as

$$s_g - s_B = -\tfrac{1}{2}m(\Delta V)^2[1/X_{Gr} + 3/X_{G(r-1)} + \ldots$$
$$+ (2n-1)/X_{G(r-n+1)} + \ldots + (2r-1)/X_{G1}] \quad (6.31)$$

and this equation can also be derived directly from Equation (6.13).

The distance $(s_g - s_B)$ can be found from Equation (6.31) if the accelerating force X_G can be estimated for any specified value of V, and hence for the mid-point of any segment in the calculation. The estimation of X_G depends on knowledge of C_D, C_L, the effective value of μ_B, the proportion of the total load carried by the braked wheels and the reversed thrust (in the speed range for which this can be used). The values used for C_D and C_L must of course allow for the effects of ground proximity and for any effects of the thrust reversers on these coefficients. Obviously, the calculation depends on the piloting technique employed and either on a large amount of empirical data or on a wide range of assumptions for a particular aircraft. A representative or instructive case would be difficult to present.

6.13 Discontinued approaches and baulked landings

For purposes of airworthiness certification it is usual to distinguish between a discontinued final landing approach, where the decision is made before the threshold, and a baulked landing where the decision to abandon the landing is made after the threshold has been passed. Williams (1972) gives a brief discussion of these two cases and lists some typical requirements for conventional transport aircraft. In either case the engine thrust will be low and the flaps will be in the landing position when the decision is made.

A decision to abandon a landing, either before or after the

threshold, may be made for various possible reasons such as the presence of wind shear, the proximity of another aircraft or a minor technical failure in the aircraft making the landing. When the decision has been made it is obviously necessary to increase the thrust as quickly as possible to the maximum in order to obtain a rapid increase of air speed, while at the same time changing the flight path from a descending glide to a climb. In principle, an adjustment of the flap angle to the lower value used for take-off would seem to be beneficial because it would reduce the drag, but in a baulked landing the safety regulations do not usually allow a change of flap angle until a satisfactory climb has been achieved at a reasonable height. Thus the primary action, on which a successful recovery is dependent, must be an increase of thrust and this can take several seconds as the turbines come up to their maximum speed.

The order of magnitude of the possible acceleration after a baulked landing can be seen by noting that in level flight the acceleration is equal to $(f - \beta)g$. For a civil transport aircraft f is likely to be about 0.25–0.3 and in the landing configuration β might be between 0.15 and 0.2 at the high value of C_L used in the approach. Thus the maximum acceleration in level flight is of order $0.1\,g$ and a climb even at an angle as small as 0.025 radian would reduce this to $0.075\,g$. These figures show that even 10 seconds after the attainment of maximum thrust the speed will have increased by only about 10 m/s if there is no climb and even less if the aircraft climbs.

6.14 The accelerate–stop distance and the balanced field length

The significance of the take-off decision speed V_1 has already been explained. If an engine failure is recognised at any speed below V_1 the aircraft is brought to rest without leaving the ground. This is the 'accelerate–stop' manoeuvre and the length required for it may be calculated in stages. First, the ground-run distance with all engines operating is calculated as explained earlier up to the speed V_{EF} at which the engine failure occurs. Then an allowance must be made for the pilot's reaction time, typically about 2 seconds, before he recognises the failure at the higher speed V_{EFR}. At this stage the pilot takes appropriate action which will include applying the brakes and spoilers, reducing the thrust of the remaining engines and operating some thrust reversal if this is available. There is then a further delay time while the mechanical systems respond to the pilot's action; proper consideration of the delays is given in ESDU 85029 (1985). The delay is usually less than 1 second for brakes and spoilers but the time for the engine thrust to fall to a low level is much greater, about 2–4 seconds if the fuel supply is cut and about 3–8 seconds if the throttle is used. The ground-run distance required to stop the aircraft is calculated using Equation (6.31). Hence the total distance required for the accelerate–stop manoeuvre can be found and plotted against the decision speed V_{EFR} at which the engine failure is recognised and action is taken to stop the aircraft. This is shown as one of the unbroken curves in Figure 6.16.

The alternative action that can be taken after failure of one engine is continuation of the take-off. The total length required for this case, up to the screen height, can be calculated by the methods already explained and this leads to the other unbroken curve in Figure 6.16. The intersection of the two curves gives the critical decision speed V_1 and the balanced field length. For $V < V_1$ the accelerate–stop action gives the shorter total length while for $V > V_1$ the continued take-off is better. The length required for $V = V_1$ is the same for either of the two alternatives and this balanced field length is the greatest distance that can be required, for all possible values of the speed at which the engine failure is recognised. Obviously, the length is very dependent on aircraft weight and other factors.

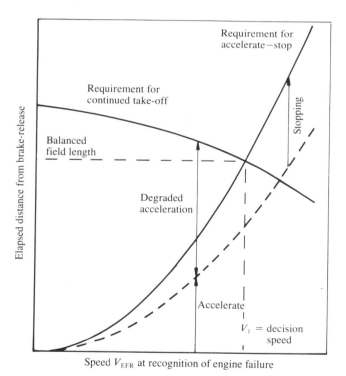

Figure 6.16. Take-off decision speed and balanced field length.

6.15 Effects of varying air temperature and pressure

Changes of air temperature and pressure have important effects on take-off and landing performance. In general, there is a reduction of engine thrust if the air pressure is reduced or the temperature increased and this reduces the acceleration during take-off and hence increases the required length of runway. A further effect of reduced thrust that is usually more important is the reduction of climb gradient after an engine failure, particularly in the critical second segment mentioned in § 6.6. Especially when the airfield is at a high altitude or the air temperature is high, the maximum permissible weight of the aircraft at take-off may be limited by the need to achieve, after the failure of one engine, the climb gradient specified in § 6.6 for the second segment of the climb. It is usual to construct for each type of aircraft a set of curves for practical use known as WAT curves, relating the maximum permissible Weight at take-off to the Altitude and air Temperature. Then, for any specified airfield with a known air temperature, the maximum weight can be found. This maximum weight, often known as the 'WAT limit', is an overriding limitation which cannot be increased by providing a longer runway, although there may of course be cases in which the length of runway restricts the allowable weight at take-off to a value below the WAT limit.

6.15.1 ENGINE CHARACTERISTICS

The effects of varying ambient air pressure and temperature on the maximum available thrust have been discussed in Chapter 5. For turbofan engines at constant air temperature,

$$F \propto p \propto \delta$$

where $\delta = p/p_0$, but the effect of varying air temperature is more complex and depends on the way in which the engine is controlled. A 'flat-rated' turbofan engine is controlled in such a way that at constant ambient air pressure there is no variation of take-off thrust with air temperature up to some specified temperature limit. Above that limit the engine operates in such a way that either (i) the fan speed or the compressor speed is constant or (ii) the turbine inlet temperature is constant and in either case the thrust then decreases with increasing air temperature. Engines which are not flat-rated are usually controlled throughout the temperature range either to keep the fan or compressor speed constant or to keep a constant turbine inlet temperature. The data given for take-off speeds in ESDU 76034 (1976) show that in the former case the thrust is roughly proportional to T^{-1} while in the latter case $F \propto T^{-2.5}$ is a fair approximation.

6.15.2 TAKE-OFF DISTANCE

For steady level flight with a given wing loading the speed V is proportional to $(\sigma C_L)^{-1/2}$. Thus the stalling speed $V_s \propto \sigma^{-1/2}$ and all the reference speeds, both for take-off and for landing, may be assumed to vary with σ in the same way. In Equation (6.5) the dominant term in the expression for the accelerating force is the thrust F, so that in Equation (6.6) the mean acceleration $\bar{a}$ is roughly proportional to F and with this approximation

$$s_R \propto (F\sigma)^{-1}.$$

The time from rotation to lift-off should not vary with F or σ and hence

$$s_{RL} \propto V_R \propto \sigma^{-1/2}.$$

The take-off airborne distance s_a is given by either of the Equations (6.22) or (6.23) and in the former case $s_a \propto \sigma^{-1/2}$ while in the latter case the variation with σ depends on the relative magnitudes of the two terms. The important effect of reduced thrust on the climb angle has already been discussed, but since the airborne part of the take-off distance is usually a relatively small proportion of the total distance, the effect of this on the total distance can be neglected. For the same reason a final consequence of the above is that the variation of the total take-off distance with F and σ can be represented by

$$(s_g + s_a) \propto F^{-m} \sigma^{-n}, \tag{6.32}$$

where each of the exponents m and n may be expected to be slightly less than 1, say between 0.8 and 1.0.

6.15.3 LANDING DISTANCE

The only effect on the landing distance caused by a variation in the maximum available thrust is due to the change in the magnitude of the reversed thrust that can be used. If this is ignored Equation (6.28) shows that $(s_g - s_B) \propto \sigma^{-1}$ and for a fixed time between touchdown and the point B shown in Figure 6.13 the distance $s_B \propto V_T \propto \sigma^{-1/2}$. The airborne landing distance is given by Equations (6.24) and (6.25). These show that the airborne distance is given by an equation of the form

$$s_a = A + B\sigma^{-1}$$

where A and B are independent of air density. The distances s_a and s_B are usually fairly small compared with $(s_g - s_B)$, so that the variation of total landing distance can be expressed as

$$(s_a + s_g) \propto \sigma^{-n},$$

Effects of wind

where n is slightly less than 1, a result very similar to that given in the relation (6.32), but acknowledging that landing distance is independent of thrust.

6.16 Effects of wind

The wind velocity near the ground may be resolved into two components, one of magnitude V_w along the runway and the other normal to the runway. Only the component V_w has any significant effect on the take-off and landing distances and this will be considered here, taking it as positive when it is a headwind. The wind velocity increases with distance above the ground and for the purpose of calculating the length of ground run required for take-off or landing it is suggested that V_w should be measured at the height of the wing, although for calculating the airborne part of the take-off or landing run a greater height should be used.

Aircraft are normally operated for take-off and landing in a direction such that the wind component V_w is positive. The wind then has a favourable effect in reducing the distances required, so that the distances are not of critical importance and approximate methods may be used for their estimation.

6.16.1 TAKE-OFF

The first distance to be considered is the take-off ground run in which the aircraft accelerates up to the air speed V_R at which rotation starts. The simplest method of estimating the effect of wind on this distance is given in ESDU EG 5/1 (1972). A factor g_w is defined as

$$g_w = (V_m - V_w)/V_m, \tag{6.33}$$

where V_m is a suitably chosen mean value of the air speed during the ground run. The factor g_w is the ratio of the mean ground speed to the mean air speed. The ground distance required in a wind is obtained by multiplying the distance calculated for zero wind by g_w. A graph relating V_m/V_R to V_w/V_R is given in ESDU EG 5/1 (1972) and this is represented with good accuracy for values of V_w/V_R up to 0.2 by

$$V_m/V_R = 0.5 + 0.27(V_w/V_R). \tag{6.34}$$

An alternative approximate method for estimating the effect of wind on the take-off ground run is also given in ESDU EG 5/1 (1972). In the ground run up to the rotation point the air speed increases from V_w to V_R, so that the distance travelled relative to the air is

$$\tfrac{1}{2}(V_R^2 - V_w^2)/\bar{a}_1$$

where $\bar{a}_1$ is a mean acceleration, calculated for a speed such that the best estimate of distance is obtained. The time taken for this increase of speed is $(V_R - V_w)/\bar{a}_2$, where $\bar{a}_2$ is a mean acceleration that is different from $\bar{a}_1$, being calculated now for a speed chosen to give the best estimate of time. The distance s_R travelled relative to the ground is calculated as

$$(\text{air distance}) - V_w(\text{time})$$

so that

$$s_R = \tfrac{1}{2}(V_R^2 - V_w^2)/\bar{a}_1 - V_w(V_R - V_w)/\bar{a}_2. \tag{6.35}$$

ESDU EG 5/1 (1972) gives values of mean air speed to be used in calculating $\bar{a}_1$ and $\bar{a}_2$. These show that for values of V_w/V_R up to about 0.2 the mean air speed should be about 0.7 V_R for $\bar{a}_1$ and about 0.6 V_R for $\bar{a}_2$, but it should be noted that these values are based on the variation of thrust with speed given by Equation (6.7) and this representation is not as good as Equation (5.8).

A better estimate of the distance s_R in the presence of a wind may be obtained by a step-by-step integration. The modified form of integral to be used is

$$s_R = \tfrac{1}{2}m \int_{V=V_w}^{V=V_R} \frac{d(V - V_w)^2}{X_G} \tag{6.36}$$

and it should be noted that in evaluating X_G for each step the true air speed V should be used, i.e. the sum of ground speed and headwind speed, not the ground speed $(V - V_w)$.

For the estimation of the effect of wind on the distances s_{RL} and s_a the factor g_w can be used, with a suitably chosen value of the mean air speed V_m for each segment.

6.16.2 LANDING

In considering the approach to landing it should be noted that it is the approach angle relative to the ground that is normally specified and controlled by the airfield guidance equipment. If this angle is γ_A the angle of glide relative to the air is $\gamma_A(1 - V_w/V_A)$ and in calculating the airborne landing distance s_a it is this new angle that should be used in Equation (6.25) instead of γ_A. Again, the factor g_w can be used to correct the calculated value of s_a for the effect of wind, using an appropriate value of V_m for the flight path from the threshold to touchdown. Taking $V_m = \tfrac{1}{2}(V_T + V_B)$ the factor g_w can also be used for estimation of the distance s_B.

For estimating the effect of wind on the landing ground run after the point B the simplest method is to use the factor g_w with $V_m = \tfrac{1}{2}V_B$. A better estimate is obtained from a modified form of Equation (6.28) viz

$$s_g - s_B = -\tfrac{1}{2}(V_B - V_w)^2/\bar{a} \tag{6.37}$$

Effects of wind

and ESDU 6/4 (1971) suggests that the mean acceleration $\bar{a}$ should be calculated for an air speed of $0.7\,V_B$, but this may be unsatisfactory when V_w/V_B is not small because of the dependence of maximum braking force on *ground* speed as indicated in Figure 6.15.

As in the case of take-off calculations, the best estimate of landing ground run in the presence of a wind can be obtained from a step-by-step integration. The integral to be evaluated is essentially the same as that in Equation (6.36), giving

$$s_g - s_B = -\tfrac{1}{2}m \int_{V=V_w}^{V=V_B} \frac{\mathrm{d}(V - V_w)^2}{X_G} \qquad (6.38)$$

but in evaluating X_G for each step it should be noted again that although the drag and thrust forces depend on V the braking force depends on the ground speed $(V - V_w)$.

7

Fuel consumption, range and endurance

The range of an aircraft is the distance that can be flown while burning a specified mass of fuel and is one of the few simple performance parameters by which the commercial value of an aircraft may be judged. Any change in design which leads to an increase in range is always desirable because it gives either

(i) a reduction in the fuel needed for a given distance, or
(ii) an increased distance for a given fuel load.

The reduced fuel of (i) reduces the cost of fuel for a specified flight and may also allow more payload to be carried for a specified total aircraft weight, provided there is the necessary space in the aircraft. The increased distance of (ii) allows the aircraft to fly over a long distance with fewer stops for refuelling and for a military aircraft it gives a valuable increase in the radius of action.

Any flight involves take-off, climb, cruise, descent and landing but except for short flights the greater part of the fuel is used in the cruise and the emphasis in this chapter is on cruising range, although the fuel used in climb and descent is also considered. The speeds and heights that should be used in the cruise to give maximum range are discussed in some detail, taking account of the constraints that are usually imposed by Air Traffic Control. It is assumed that the aircraft is flying in still air, except in § 7.12 where it is shown that a wind affects not only the range relative to the ground but also the optimum speed for achieving maximum ground range.

The endurance of an aircraft is the time for which flight can be maintained while burning a specified mass of fuel. This is usually much less important than range, but long endurance is required for some special purposes such as ocean patrol and the characteristics which lead to long endurance will be beneficial for any aircraft while holding in a stack, awaiting permission to land. The speeds and heights required for maximum endurance are considered in this chapter.

7.1 The phases of a flight

In any flight, fuel is used from the moment when the engines are first started until they are shut down finally after landing at the destination. The phases to be considered in a normal flight, without diversion are:

- (i) Start engines, taxi to end of runway and wait for take-off.
- (ii) Take-off and initial climb (say to 500 m).
- (iii) Climb to height required for start of cruise.
- (iv) Cruise.
- (v) Descend to height required for start of landing approach.
- (vi) Decelerate to approach speed.
- (vii) Approach and land.
- (viii) Taxi from runway to unloading point.

Phases (i) and (viii) obviously make no useful contribution to the distance flown and it is usually assumed that this is also true for phases (ii), (vi) and (vii), although the fuel used in all these phases must be included in the estimate of total fuel consumption. The effects of phases (iii) and (v) on the range and time of flight are most easily estimated by calculating the 'lost range' and 'lost time' for each phase. The lost range for the climb is found by estimating the mass of fuel used and the distance flown in this phase and then calculating the distance that would be flown in the cruising condition while using this mass of fuel. The difference between the latter distance and the actual climb distance is the lost range in the climb. Similar calculation procedures are used for the descent and for finding the lost time in climb and in descent. The calculations are explained fully in § 7.10.

7.2 Fuel reserve and allowances

The greater part of the *reserve* fuel is carried to allow diversion to an alternative airfield if for any reason the aircraft cannot land at the planned destination. In calculating the mass of fuel needed for such a diversion, it is assumed that the aircraft descends to ground level at the planned destination but is then refused an approach, overshoots and climbs to a height in the region of 7 or 8 km. There is then a short cruise at this height before descending to the alternative airfield which is usually assumed to be within about 350 km of the planned destination. It is also assumed that the aircraft will be required to hold in a stack at a height of about 1500 m for a period up to 45 minutes in the vicinity of the alternative airfield. If diversion is not required, this part of the reserve fuel is available for holding in a stack near the planned destination, but the reserve is sometimes required to include fuel for two holding periods, one near the planned destination and the other near the alternative airfield.

164 *Fuel consumption, range and endurance*

An additional fuel reserve may also be needed to allow for the possible failure of one engine. This reduces the maximum possible height for cruising and may require significant extra fuel, as discussed in § 7.11.

Whereas the reserve fuel is not normally used the fuel *allowances* are expected to be used and are intended to cover:

(a) Minor deviations from the planned flight path.
(b) Departures from the planned cruising speed and height which may occur because of restrictions imposed by Air Traffic Control (ATC) or for other reasons.
(c) Fuel used in phases (i), (ii), (vi), (vii) and (viii) as listed in § 7.1.

For (a) and (b) it may be sufficient simply to allow 5% of the 'stage fuel', i.e. the fuel that is actually used in a normal flight. For phases (i), (vi), (vii) and (viii) as listed in § 7.1 the engines may be assumed to be at very low thrust, close to the idling condition, thus requiring, say, 2–3% of the stage fuel. For phase (ii) it is usual to assume that take-off thrust is used for a period of $1\frac{1}{2}$ minutes, this giving a sufficiently accurate assessment of the fuel that is needed.

The fuel used in the cruising phase of the flight can be a surprisingly small fraction of the total fuel load carried at the start, especially when the stage length is short. This is illustrated by the figures in Table 7.1, which refer to a typical aircraft operating over a stage length of 1000 km and give the mass of fuel used in each phase as a percentage of the total fuel carried at the start of the take-off. In this case the fuel used in the cruise is only 22.8% of the total and even the sum of climb, cruise and descent fuel is only 48.2% of the total, whereas the reserve fuel required for diversion is nearly 40%. For a longer stage length the proportion used in the cruise would of course be greater and the reserve proportion would be smaller.

7.3   Work done for a specified range

Before discussing cruising range in detail it is instructive to consider a very simple approach to show the effects of increasing height on the work done and fuel used in moving an aircraft over a specified range. Neglecting the change of weight due to use of fuel and considering only the horizontal portion of the flight path, C_L and β remain constant if the speed is constant. The drag is βW and the work done in moving the aircraft over a range R_a is $\beta W R_a$. This has the minimum value $\beta_m W R_a$ when $V_e = V_e^*$. Provided the height is never so great that the Mach number corresponding to V_e^* becomes high enough to cause a significant rise of C_D, the minimum work is independent of height and an increased height is beneficial because it

Table 7.1. *Fuel used by a typical short-range civil transport with turbofan engines.*

(a) *Normal flight, without diversion*	
Start-up and taxi	1.29
Take-off	3.25
Climb	23.33
Cruise	22.80
Descent	2.07
Approach	1.44
Taxi	1.29
Flight contingencies (allowances)	4.66
	60.13
(b) *Diversion using reserve fuel*	
Overshoot	2.50
Climb	12.25
Cruise	6.60
Descent	1.60
Hold	15.33
Approach	1.59
	39.87

The figures shown are percentages of the total fuel load at the start of the take-off. Stage length = 1000 km.

gives a higher true air speed for the same minimum work. As an example, a comparison may be made between flight at sea level and at a height of 12 km, where $\sigma \approx 0.25$. For $V_e = V_e^*$ the drag and thrust are the same at the two heights, so that the work for a given range is the same, but the speed at $h = 12$ km is twice that at sea level.

For a turbofan engine with a specific fuel consumption $c = Q/F$ the ratio of work done to mass of fuel used is V/c and this is proportional to the overall efficiency of the engine. If c is given by Equation (5.11),

$$V/c = V/(c_1 \theta^{1/2} M^n)$$

and since $V = Ma_0 \theta^{1/2}$ this is proportional to $M^{(1-n)}$. Table 5.1 shows that the exponent n is always less than 1, so that with constant V_e the ratio V/c increases with height. This means that for an aircraft powered by turbofan engines an increase of height gives both an increase of speed and a reduction of fuel consumption for the minimum work condition, and it is essentially for this reason that the modern jet transport can offer high efficiency with high speed, but only if it is allowed to fly high.

For a propeller-driven aircraft increasing height gives the same advantage of increasing speed, for the minimum work condition, but the reduction of fuel consumption is smaller than it is for a turbofan aircraft. This can be seen from the account in § 5.4.2 of the variation of the specific fuel consumption $c' = Q/P_e$ with speed and height. The matter will be discussed further in § 7.4.

7.4 Basic equations for cruise range

For an aircraft powered by turbojet or turbofan engines the specific fuel consumption is defined as $c = Q/F$, where Q is the mass of fuel used per unit time and F is the thrust. In level cruising flight

$$F = D = \beta L = \beta W$$

and the mass of fuel used per unit *distance* is

$$cF/V = c\beta W/V.$$

As the aircraft travels a distance dR_a (in still air) its mass changes from m to $(m + dm)$ where

$$dm = -(c\beta W/V) \, dR_a.$$

The *specific range* r_a is defined as the distance travelled per unit mass of fuel used. Thus

$$r_a = -dR_a/dm = V/(c\beta W). \tag{7.1}$$

A measure of the overall efficiency of the aircraft as a load carrier is

$$Wr_a = \frac{\text{Weight} \times \text{Distance}}{\text{Mass of fuel used}}$$

and this is equal to $V/(c\beta)$, the product of the aerodynamic efficiency L/D and the ratio V/c which is proportional to the overall efficiency of the engine and which increases with speed, as explained earlier.

Putting $W = mg$, Equation (7.1) shows that the total cruise range is

$$R_{ac} = -\int_{m_1}^{m_2} r_a \, dm = -\int_{m_1}^{m_2} \frac{V}{gc\beta} \frac{dm}{m}, \tag{7.2}$$

where m_1 and m_2 are the masses of the aircraft at the beginning and end of the cruise. In general this integral can be evaluated by dividing the total mass of fuel used $(m_1 - m_2)$ into equal stages and calculating r_a for each stage, based on a value at the mid-point, although for flights that are not very long it is often sufficiently accurate to treat the whole cruising flight as one stage. Then the range is given by the product of the mean value of r_a and the total mass of fuel used.

Basic equations for cruise range

It will be shown later that in the cruise procedure that leads to maximum range, the quantities V, c and β are either constant or nearly so. If they are all exactly constant the evaluation of the integral in Equation (7.2) is particularly simple and

$$R_{ac} = [V/(gc\beta)] \ln(m_1/m_2). \tag{7.3}$$

This is known as the *Breguet range equation* and the range calculated from it, with specified values of V, c and β, is often known as the *Breguet range*.

The Equation (7.3) embodies an unusually concise, if implicit, statement about design efficiency, for the three principal factors c, β and the mass ratio display by their values a measure of success in propulsion, aerodynamic and structural design respectively. The first of these, specific fuel consumption, indicates how small a fuel flow rate is required to produce a unit of thrust and much effort has been devoted to reduction of that quantity. The second factor, $\beta = C_D/C_L$, is a measure of the aerodynamic efficiency of the aircraft and is likewise minimised as far as possible, while the mass ratio is related to structural efficiency as explained below.

The mass m_2 at the end of the cruise is, in the extreme case, close to the zero-fuel mass of the aircraft and comprises the structure, power plant, systems and payload. The mass m_1 at the start, again in the extreme case, is the maximum allowable total mass after loading the payload and fuel. If it is assumed that for a given mission the necessary masses of payload, systems and power plant are fixed, the remaining mass making up m_2 is that of the structure necessary to contain and support not only the payload, systems and power plant, but also the fuel.

The mass of fuel used is

$$m_f = m_1 - m_2 \tag{7.4}$$

and the mass ratio is

$$m_1/m_2 = 1 + m_f/m_2. \tag{7.5}$$

Equation (7.3) shows that long range requires a large value of m_1/m_2 and Equation (7.5) shows that this implies that m_f/m_2 should be large. Thus a large fuel mass and small m_2 are needed and in order to achieve this the structure mass should be small, i.e. the structure should be able to support a large multiple of its own weight.

A typical turbofan transport might have a structure mass of 20 000 kg, i.e. without the power plant, systems, payload and fuel. The zero-fuel mass m_2 (with the payload, systems and power plant) might be 55 000 kg and the maximum take-off mass m_1 might be 70 000 kg. The extreme value of the mass ratio m_1/m_2 would then be 1.273, but if a more efficient structure were used, having 10% lower mass while still allowing the same maximum take-off mass, the ratio m_1/m_2 for the same payload would become 1.321 and the Breguet range would be increased by 15%.

EXAMPLE 7.1. BREGUET RANGE OF TURBOFAN AIRCRAFT

An aircraft cruises at a constant value of $V/(c\beta)$ equal to 2.3×10^8 (m/s)^2. If the mass at the start of the cruise is 100 000 kg and the mass of fuel used is 25 000 kg, find the cruise range (a) from Equation (7.3) and (b) by using a single mean value of the specific range.

The common factor required for both calculations is

$$V/(gc\beta) = (2.3 \times 10^8)/9.81 = 2.345 \times 10^7 \text{ m}$$

and the direct use of this in Equation (7.3) gives

$$R_{ac} = (2.345 \times 10^4) \ln(100/75) \text{ km} = 6745 \text{ km}.$$

For the simpler approach the mean value of $1/m$ is used, leading to

mean specific range
$$= \tfrac{1}{2}(2.345 \times 10^7)(1 + 4/3) \times 10^{-5} \text{ m/kg}$$
$$= 273.58 \text{ m/kg},$$

from which the cruise range is found to be

$$R_{ac} = 273.58 \times 25 \text{ km} = 6840 \text{ km}.$$

The difference between the two values of R_{ac} is less than 1.5%.

For an aircraft powered by turboprops the specific fuel consumption is usually defined as $c' = Q/P_e$ and the equations for range and specific range then take a different form. The thrust is

$$F = \eta_{PR} P_e/V = \beta W$$

and the mass of fuel used per unit distance is

$$c' P_e/V = c' \beta W/\eta_{PR}.$$

The specific range is then

$$r_a = -dR_a/dm = \eta_{PR}/(c'\beta W) \tag{7.6}$$

and the total cruise range is

$$R_{ac} = -\int_{m_1}^{m_2} \frac{\eta_{PR}}{gc'\beta} \frac{dm}{m}. \tag{7.7}$$

In contrast to Equation (7.2) the speed V does not appear in the numerator of the integrand and for a given value of β the range does not increase with speed unless c'/η_{PR} decreases. Changes of η_{PR} with speed and height are usually fairly small and as explained in § 5.4.2 there is only a small decrease of c' with increasing speed. At the cruise rated power there is little variation of c' with height but at a constant level of power below the rated value there is usually some decrease of c' as height is gained, because of the increase in the ratio of the actual shaft power P_s to the rated power P_{sm}. Thus for this class of aircraft an increase of height, with its associated increase of true air speed for minimum β, does give some increase of range, but as noted in § 7.3 the gain is smaller than it is for a turbofan aircraft.

For a turboprop aircraft with constant values of η_{PR}, c' and β during the cruise, Equation (7.7) gives

$$R_{ac} = [\eta_{PR}/(gc'\beta)] \ln(m_1/m_2) \qquad (7.8)$$

and this is the appropriate form of the Breguet range equation for this class of aircraft. As for the turbofan aircraft the range calculated from the equation is known as the Breguet range and in accordance with the earlier discussion of Equation (7.3) this form of the Breguet range equation also demonstrates the influence of three measures of efficiency on range.

7.5 Conditions for maximum cruise range – turbofans

Equation (7.2) shows that the cruise range R_{ac} is maximum when the specific range r_a has the maximum value at every instant of the flight. (It should be noted that r_a depends on the instantaneous conditions of flight and is potentially changeable during the flight.) The requirement for maximum R_{ac} is therefore that $c\beta/V$ should always have the minimum possible value. As fuel is used the weight W falls and it will be shown that either a gradual climb or a reduction of speed is needed in order to keep $c\beta/V$ at its minimum value. The conditions required during the cruise depend on what is kept constant (e.g. h, V or M) and on whether

  (a) the air temperature is constant ($h > 11$ km in the ISA),
  (b) the specific fuel consumption c may be assumed to be constant or should be represented by Equation (5.11), or
  (c) the conditions are limited by the available thrust.

Various cases will be considered and for each of these the optimum value of the speed ratio v will be found, enabling the appropriate combination of speed and height to be calculated from the specified conditions for any value of the weight. The range R_{ac} can then be found from Equation (7.2), or from Equation (7.3) if $V/(gc\beta)$ remains constant during the cruise. Most of the results are summarised in ESDU 73019 (1982).

7.5.1 CONSTANT TRUE AIR SPEED

If the true air speed V has a specified constant value the condition for minimum $c\beta/V$ is that $c\beta$ should be minimum. In the ISA with $h > 11$ km both the temperature and the Mach number are constant and Equation (5.11) shows that c is also constant. Thus minimum β is required, i.e. $v = 1$, $V_e = V_e^*$ and C_L must have the constant value C_L^*. Since $C_L = w/(\frac{1}{2}\rho_0\sigma V^2)$ the required constant value of C_L can be achieved only by gradually increasing the height as fuel is used so that $\sigma \propto W$. This form of cruise is known as a *cruise-climb* and Equation (7.3) shows that it gives a total cruise range

$$R_{ac} = [V/(gc\beta_m)] \ln(m_1/m_2). \tag{7.9}$$

To achieve the required conditions, the cruising height at any given weight must be adjusted so that $\sigma = (V_e^*/V)^2$, where V is the specified true speed. The range R_{ac} as given by Equation (7.9) appears to increase indefinitely with V (i.e. with height), but the equation is valid only if the Mach number is not much greater than the drag divergence value M_D defined in § 2.4, so that there is no significant increase in C_D and β_m due to rising Mach number. Equation (7.9) shows that maximum cruise range is obtained when the specified constant value of V is chosen to give maximum $V/(c\beta_m)$ and this value of V will be close to $M_D a$, where a is the velocity of sound. In some cases the available thrust may be insufficient to enable the aircraft to fly at $v = 1$ with V determined in this way. The best cruising conditions are then limited by the available thrust and this will be considered in § 7.5.4.

The equations given so far in this chapter have all been based on the balance of forces in level flight, i.e. $F = D$ and $L = W$. If γ is the mean angle of climb in radians during the cruise, Equation (2.3) shows that the effect of the small angle of climb is equivalent to increasing β to $(\beta + \gamma)$, so that the range is reduced in the ratio $\beta/(\beta + \gamma)$. This correction can easily be applied if high accuracy is required but it is always small, as illustrated in the following example.

EXAMPLE 7.2. EFFECT OF CLIMB ANGLE IN CRUISE-CLIMB

The aircraft considered in Example 7.1 cruises with $\beta = \beta_m = 0.06$ and starts the cruise at a height of 11 km. Find the corrected range, allowing for the angle of climb.

Keeping $\sigma \propto W$ in order to maintain constant C_L, the ratio of densities at the end and start of the cruise must be

$$\sigma_2/\sigma_1 = m_2/m_1 = 0.75$$

and so

$$\sigma_2 = 0.75 \times 0.2971 = 0.2228,$$

Conditions for maximum cruise range – turbofans

for which the ISA conditions give

$$h_2 = 12.84 \text{ km}.$$

Then the mean angle of climb is

$$\gamma = 1.84/6745 = 2.73 \times 10^{-4} \text{ radians}$$

and the corrected range is

$$R_{ac} = \frac{0.06}{0.06 + (2.73 \times 10^{-4})} \times 6745 = 6714 \text{ km}.$$

The reduction of range is only about 0.5%.

If the temperature decreases with increasing height as in the ISA with $h < 11$ km, there will be an increase of Mach number in a cruise at constant V because of the reduction in the velocity of sound as the height increases. Then if M is not to be much greater than M_D at the end of the cruise it will usually be below that value at the start. If it is sufficiently accurate to assume that c is constant, even though θ and M vary to some extent, the condition required for maximum specific range is still $\beta = \beta_m$, $V_e = V_e^*$, $\sigma = (V_e^*/V)^2$ and $\sigma \propto W$, but it is more realistic to represent the specific fuel consumption by Equation (5.11).
Then, since

$$M = V/a = V/(a_0 \theta^{1/2}),$$
$$c = c_1 \theta^{1/2} M^n = c_1 (V/a_0)^n \theta^{(1-n)/2}. \tag{7.10}$$

For maximum specific range $c\beta$ is required to be minimum and this implies minimum $\beta \theta^{(1-n)/2}$, a result that is independent of the relation between C_D and C_L. Equation (1.6) shows that in the ISA with $h < 11$ km

$$\sigma = \theta^{(m-1)}, \tag{7.11}$$

where $m = 5.256$ as shown in Appendix 2. Assuming that the simple parabolic drag law is valid and using these results with Equation (3.18), it can be shown that the speed ratio required to give minimum $c\beta$ is

$$v = [(7.512 + n)/(9.512 - n)]^{1/4}. \tag{7.12}$$

Table 5.1 shows that typical values of n are in the range 0.2–0.5, for which v varies between about 0.95 and 0.97. The corresponding values of β/β_m are only slightly greater than 1, about 1.003 for $v = 0.96$.

When $h < 11$ km, with either of the two assumptions about the specific fuel consumption c, maximum range in a cruise at constant V will be obtained if sufficient thrust is available by keeping v at the optimum value and choosing V so that at the end of the cruise the Mach number is close to M_D.

7.5.2 CONSTANT MACH NUMBER

When $h > 11$ km the velocity of sound is constant and the condition of constant M is the same as that of constant V which has already been considered. When $h < 11$ km Equation (5.11) gives $c \propto \theta^{1/2}$ because M is constant. The condition of minimum $c\beta/V$ then requires minimum $\theta^{1/2}\beta/(a_0\theta^{1/2}M)$, i.e. minimum β. Thus the condition for maximum range in this case is $\beta = \beta_m$, i.e. $v = 1$ as for the cruise with constant M and V when h is greater than 11 km. Again, the height should be increased gradually to keep $\sigma \propto W$ and if sufficient thrust is available the specified cruise Mach number should be chosen as discussed in § 7.5.1 for $h > 11$ km, so that it is close to M_D and $V/(c\beta_m)$ has the maximum value.

If c is assumed to be constant, for this case of constant M with $h < 11$ km, the condition for minimum $c\beta/V$ is that $\beta/(Ma_0\theta^{1/2})$ should be minimum and this requires minimum $\beta/\theta^{1/2}$. This gives a speed ratio v that is rather greater than 1, because the height for the given M is then less than that giving $V_e = V_e^*$ and this gives a greater value of θ. Since the assumption of constant c is less realistic than the use of Equation (5.11) and the analysis is more complex with this assumption, it will not be pursued here.

7.5.3 THRUST ADJUSTMENTS IN A CRUISE-CLIMB

In a cruise-climb with $h > 11$ km the temperature is constant and if V, M and the engine speed N are all kept constant the functional relation (5.5) shows that

$$F \propto \delta \propto \sigma.$$

Also, if β is constant,

$$D \propto W \propto \sigma$$

and hence the ratio F/D is unchanged, so that if the engine speed is set initially to give the required thrust, no further adjustment will be needed during the cruise.

With $h < 11$ km the temperature falls as the aircraft climbs and as explained in § 5.2.2 this causes the ratios F/δ and F/σ to increase, if the Mach number and engine speed are kept constant. Since the drag is still proportional to σ this means that the engine speed must be reduced gradually as the climb proceeds, until the height of 11 km is reached. This is still true if V is constant instead of M, because the variation of Mach number in a climb at constant V is small and has little effect on the thrust.

7.5.4 SPEED AND HEIGHT LIMITED BY AVAILABLE THRUST

It has been assumed so far that sufficient thrust is available to enable the aircraft to cruise at the speed and height required ideally to give maximum range. In many cases this is not a realistic assumption and it will be shown later that when the maximum engine thrust (and thus the engine weight) is taken into account the optimum conditions are obtained when the cruising speed and height are limited by the available thrust, i.e. the optimum is defined at maximum thrust rather than minimum β. This situation will now be considered, initially subject to the simplifying assumption that c is constant, but later allowing for a realistic change of c. The conditions will be examined for heights both above and below 11 km in the ISA, the objective being the same in both cases, namely to consider what can be achieved when an upper limit of F is imposed and whether there is an advantage in flying at $\beta = \beta_m$. For all the cases considered the speed ratio v for best range will be found, this implying an associated value for β which is not necessarily the minimum value β_m.

The initial case to be considered rests on the assumption that the maximum available (cruising) thrust is always used. In the ISA for $h > 11$ km and with constant N and M,

$$F \propto \delta \propto \sigma$$

as noted in § 7.5.3, and this is still nearly correct even if there is some small variation of M. Then since

$$F = D = \tfrac{1}{2}\rho_0 \sigma V^2 S C_D,$$

the speed can be expressed as

$$V = [2F/(\rho_0 \sigma S C_D)]^{1/2}. \tag{7.13}$$

The condition for maximum range is that $c\beta/V$ should be minimum and if c is assumed to be constant this requires minimum β/V. Then if F/σ is constant (the engine operating at its rated cruise thrust), Equation (7.13) shows that

$$V \propto C_D^{-1/2}$$

so that

$$\beta/V \propto C_D^{3/2}/C_L.$$

The quantity that is required to be minimum is therefore

$$C_D/C_L^{2/3} = \beta C_L^{1/3}$$

and since $C_L \propto V_e^{-2}$ the requirement is that $\beta V_e^{-2/3}$ should be minimum. This conclusion is independent of the relation between C_D and

C_L but if the simple parabolic drag law is assumed to be valid, Equation (3.18) shows that for a given weight, i.e. given V_e^*,

$$\beta V_e^{-2/3} \propto (v^{4/3} + v^{-8/3}).$$

The condition for this to be minimum is that

$$v = 2^{1/4} = 1.189 \tag{7.14}$$

and this leads to $\beta/\beta_m = 1.061$, the associated condition for maximum range.

It may seem surprising that the optimum speed ratio is nearly 20% greater than that for minimum β. In particular, V_e^2 is more than 40% greater than the value V_e^{*2} for $\beta = \beta_m$, yet the drag force as given by βW is only 6% greater. The explanation lies in the greatly reduced value of C_D, due to the reduction of C_L associated with the increased value of V_e. With constant F/σ, the increase of F required to balance the extra drag is obtained by a reduction of height, requiring 6% increase of σ. The true air speed is 15% above the value for $\beta = \beta_m$ at the reduced height and β/V is about 8% lower.

If c is not assumed to be constant but is represented by Equation (5.11), minimum $c\beta/V$ is obtained when $\beta V^{(n-1)}$ is minimum, because when $h > 11$ km the temperature is constant and $M \propto V$. Since Equation (7.13) gives $V \propto C_D^{-1/2}$ the condition required is minimum $C_D^{(3-n)/2}/C_L$ and it can be shown from Equation (3.18) that the corresponding speed ratio is

$$v = (2-n)^{1/4}. \tag{7.15}$$

For the usual range of n this gives values of v ranging from about 1.08 (for $n = 0.64$) to the same value as given by Equation (7.14) for $n = 0$.

The results obtained for this case, whether c is assumed to be constant or not, show that in order to achieve the best range v, β and C_L are all required to be constant and therefore a cruise-climb is needed to keep $\sigma \propto W$, as in the other cases considered earlier. Also $V_e \propto V_e^* \propto \sigma^{1/2}$, so that the true air speed V and the Mach number are constant, and at the rated cruise thrust $F \propto \sigma \propto W$.

When $h < 11$ km, increasing height gives a reduction of temperature and an increase of F/σ. This has been discussed in § 5.2.2, where it was shown that the increase may be represented by a power law

$$F/\sigma \propto \sigma^{-r}$$

where the value of r for a typical civil turbofan is about 0.4. If c is assumed to be constant the condition for maximum range is that β/V is minimum and it can be shown that

$$\beta/V \propto \sigma^{r/2} C_D^{3/2}/C_L.$$

Noting as before that $C_L \propto V_e^{-2}$ and using Equation (3.18) it can be shown that the required value of the speed ratio is

$$v = [(4-3r)/(2-r)]^{1/4} \qquad (7.16)$$

and with $r = 0.4$ this gives $v = 1.15$, slightly higher than the mean of the values given by Equation (7.15).

There are some further points to be discussed in relation to cruise-climb procedures where the available thrust may be a limiting factor. These are most easily explained by considering a particular example and for simplicity it will be assumed that the specific fuel consumption c is constant and that $h > 11$ km, so that the temperature is constant. In Figure 7.1 the broken and full lines represent true air speed and specific range respectively and these are plotted against v for an aircraft of 65 000 kg mass, with $c = 2 \times 10^{-5}$ kg/N s, $S = 150$ m²

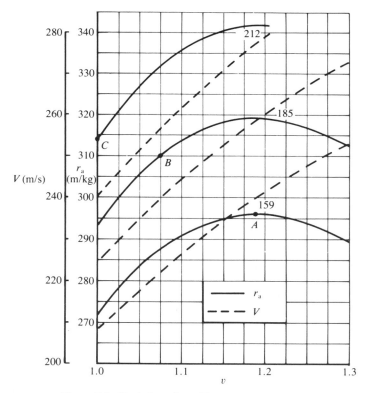

Figure 7.1. Variation of specific range with speed ratio and thrust – turbofan aircraft with cruise conditions limited by available thrust. $c = 2 \times 10^{-5}$ kg/N s, $m = 65\,000$ kg, $S = 150$ m², $K_1 = 0.02$, $K_2 = 0.045$. Numbers on the curves are values of F/σ in kN.

and drag characteristics based on $K_1 = 0.02$ and $K_2 = 0.045$. The curves have been derived by calculating C_L and β from values of C_L^*, β_m and v, then replacing C_D in Equation (7.13) by βC_L and finding V. Results are shown for three different engines having different values of F/σ for the rated cruise thrust and it is assumed that for speeds above 240 m/s ($M > 0.813$) the increase of C_D and β with M would make the aircraft uneconomic and the calculations unrealistic. For all three values of F/σ the calculated specific range r_a is maximum at $v = 1.19$ in accordance with Equation (7.14), but for the two higher values of F/σ the speeds for this optimum condition are 259 and 277 m/s, so that they are above the assumed limit of 240 m/s. The limiting case occurs when F/σ is 159 kN; the maximum specific range can then be obtained (at the point A on the curve) with the optimum value $v = 1.19$ and with V just at the limiting value of 240 m/s. For this case and for all smaller values of F/σ the cruise conditions for maximum r_a are limited by the available thrust and are obtained with the same value of $v = 1.19$ and with $V \leq 240$ m/s. The values of v and β for maximum r_a remain constant as F/σ is reduced below the limiting value of 159 kN and since $D = \beta W$ the thrust required for a given weight also remains constant and is obtained with smaller F/σ by reducing the height. Provided that the height remains greater than 11 km (so that the temperature is constant and for a given engine F/σ is independent of height), Equations (7.1) and (7.13) show that the maximum value of r_a varies in direct proportion to $(F/\sigma)^{1/2}$, and the three maxima shown in Figure 7.1 display this dependence.

As F/σ is increased above 159 kN the imposed speed limit of 240 m/s forces a decrease in r_a below the maximum and thus requires a reduction of v below the optimum value of 1.19. Thus the points B and C both correspond to $V = 240$ m/s for the larger values of F/σ, but v is progressively reduced as F/σ rises until, for $F/\sigma = 212$ kN, the maximum range obtainable within the constraint imposed by transonic drag rise is available at $v = 1.0$ (and hence $\beta = \beta_m$), with no limitation imposed by the available thrust.

To summarise the conclusions for the particular aircraft represented in Figure 7.1, a cruise-climb at $\beta = \beta_m$ (i.e. $v = 1.0$), with no limitation imposed by thrust but restricted to a maximum speed of 240 m/s by the drag-rise criterion, requires that F/σ be 212 kN. However, if F/σ is limited to a lower value, while retaining the same speed restriction, the best specific range is obtained at a value of v greater than 1.0 but less than the ideal value of 1.19, until F/σ falls to 159 kN. Then r_a is at its peak for that value of F/σ, although it has been reduced at every reduction of F/σ as shown by the points C, B, A. As F/σ is reduced below 159 kN the optimum value of v remains at 1.19 but the maximum value of r_a continues to fall.

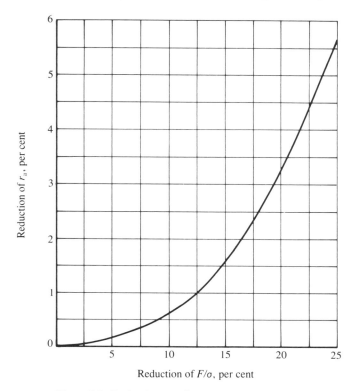

Figure 7.2. Reduction of F/σ and r_a from point C in Figure 7.1.

Figure 7.2 refers to the same aircraft as Figure 7.1 and shows the relation between the percentage reductions of F/σ and r_a, enabling the trade-off between installed engine power and range to be evaluated. The figure shows how little the range is reduced as F/σ falls from 212 kN, the value required for a cruise at $\beta = \beta_m$, while maintaining the limitation that V must not exceed 240 m/s. Since engine weight is roughly proportional to F/σ, there is always likely to be some benefit in reducing F/σ and departing from the apparent ideal of $\beta = \beta_m$, thus allowing the optimum cruise conditions to be determined by the limited thrust. For example, a decision to reduce F/σ, and hence the engine weight, by 10% would cause the specific range to fall by only 0.6%.

7.5.5 CONSTANT HEIGHT

The cruise-climb procedures discussed in § 7.5.1–7.5.4 are not usually allowed by Air Traffic Control (ATC), except in air space with very little traffic. Civil aircraft are usually required by ATC to cruise at a constant height, although a *stepped cruise* is sometimes allowed on a long flight. In this form of cruise the aircraft flies at a constant height for a specified distance and then climbs in a 'step' to a new height where again it flies at a constant height. This stepping procedure may be repeated several times but there are usually only a few steps and the cruise is then best considered as a close approximation to a cruise-climb at a specified constant true air speed or Mach number.

For the more realistic case of a cruise at a specified constant height, two cases will be considered, one with c constant and the other with c varying in accordance with Equation (5.11). If c is assumed to be constant, Equation (7.1) shows that the condition for maximum specific range at any given weight is that β/V should be minimum, and since $V_e \propto C_L^{-1/2}$ this means that $C_D/C_L^{1/2}$ should be minimum, whatever the relation between C_D and C_L. If the simple parabolic drag law is valid Equation (3.18) shows that for a given height

$$\beta/V \propto \beta/V_e \propto (v + v^{-3})$$

and the condition for this to be minimum is

$$v = 3^{1/4} = 1.316. \tag{7.17}$$

If c is not constant but varies with speed, the condition for maximum specific range is that $c\beta/V$ should be minimum and it can be shown from Equations (3.18) and (5.11) that the condition leads to

$$v = [(3-n)/(1+n)]^{1/4}. \tag{7.18}$$

For the usual range of n as given in Table 5.1 this gives values of v between about 1.1 and 1.25.

At constant height the optimum value of v given by either of the Equations (7.17) or (7.18) can be maintained only if the speed is decreased as fuel is used, because $V_e = vV_e^*$ and $V_e^* \propto W^{1/2}$. This decrease of speed is not usually acceptable and it is necessary to consider a cruise procedure that is an alternative to constant v. Figure 7.3 is based on a form of plotting used by Peckham (1970) and shows for a typical civil aircraft the relations between specific range, speed, aircraft mass and the speed ratio v. The characteristics of the aircraft are the same as those assumed for Figure 7.1, but a substantial variation of mass has been introduced and again the simple assumption has been made that c is constant. The height of 8 km has been chosen to allow all the essential features of the diagram to be made clear without including unacceptable Mach numbers greater than about 0.8. The choice of a rather greater height would have increased all the true air speeds and would have given an increase of range, especially for the case of constant speed, but this will be discussed later.

Conditions for maximum cruise range – turbofans

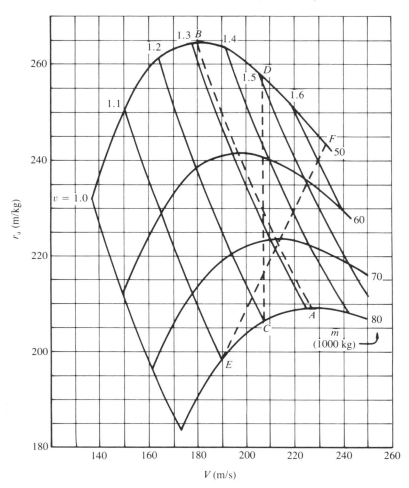

Figure 7.3. Variation of specific range with speed and aircraft mass at constant height – turbofan aircraft. $c = 2 \times 10^{-5}$ kg/N s, $h = 8$ km, $S = 150$ m^2, $K_1 = 0.02$, $K_2 = 0.045$.

The aircraft represented in Figure 7.3 has a mass of 80 000 kg at the start of the cruise and 50 000 kg at the end. Thus the ratio of the mass of fuel used to the initial mass is 3/8 and for such a high value of this ratio it would be unusual to cruise for the whole distance at constant height, without any steps to greater height. The rather large ratio of fuel to aircraft mass is assumed here in order to illustrate clearly the effects of varying total mass during a cruise at a constant specified height.

Figure 7.3 shows that for any given mass the specific range is maximum at $v = 1.316$ in accordance with Equation (7.17). As mentioned earlier, a cruise maintaining this optimum condition involves a reduction of speed as fuel is used, so that $V \propto W^{1/2}$. This is shown by the broken line AB in Figure 7.3, where the ratio of final to initial speed is seen to be $180/227.5 = (5/8)^{1/2}$. This procedure, keeping v constant, is not usually acceptable because the height must be low enough to ensure that the Mach number at the start is not much greater than M_D and unless the range is short the speed at the end of the flight is always considerably less than the initial value. Thus the average speed is usually unacceptably small and lower than that which is obtainable by using the next procedure to be considered.

The vertical broken line CD in Figure 7.3 shows a cruise at constant speed, where v necessarily rises as the weight falls, because $V_e^* \propto W^{1/2}$. The problem then arises as to how the height and speed should be chosen to maximise the range. The greatest range is obtained by making the mean value of v near to the optimum; v is then below the optimum value in the early part of the cruise and above it in the later part, as indicated by CD in Figure 7.3, but it is also important to keep the Mach number close to M_D to obtain the advantages of higher speed and greater range.

A mean value of v can be selected, near to the optimum of 1.316, by arranging for the vertical line CD to cross the locus of maxima AB near the mid-point of the mass range, i.e. at about 65 000 kg in Figure 7.3. The remaining task is then to prepare the set of curves for a height which puts CD at a value of V corresponding to the desired limiting Mach number. In the particular case represented in Figure 7.3, the Mach number for the line CD is only about 0.67, but an estimate can be made of the height which should be used in order to bring the Mach number up to the limiting value of 0.813 which was chosen for Figure 7.1. The speed ratio v for any given mass is determined by the end points C and D and if c is assumed to be constant both the true speed and the specific range for the specified mass are proportional to $\sigma^{-1/2}$; thus they both increase with height. The desired increase of M (and thus V) is in the ratio $0.813/0.67 = 1.21$, indicating that σ should be reduced in the ratio $(1.21)^{-2} = 0.68$, and this can be achieved by increasing the height from 8 km to about 11.1 km. The specific range for any specified mass is then also increased in the ratio 1.21 and the mean value of about 232 m/kg shown for the line CD in Figure 7.3 is increased to 281 m/kg.

It should be emphasised that Equations (7.17) and (7.18) are valid only when the height has a specified constant value but the speed is allowed to vary. If the aircraft is required to cruise at a specified constant speed or Mach number the analysis of § 7.5.1 or 7.5.2 should be used instead. The optimum flight path is then a cruise-climb with

$v \approx 1$, but if flight at a constant height is required this height should be chosen to keep the mean value of v as close as possible to 1 or to the value given by Equation (7.12), where this is relevant. The distinction between the cases of specified height and specified speed is illustrated further in Example 7.3 in § 7.8.

The cruise procedures represented by the lines AB and CD in Figure 7.3 both require a reduction of engine thrust as the weight falls. For the line AB, with constant v, the drag/lift ratio β is constant and since $F = D = \beta W$ the required thrust F must fall in proportion to W. For the line CD, with constant V, the lift coefficient C_L must be reduced in proportion to W; hence C_D and the drag also fall and the thrust must be reduced.

It is sometimes convenient to keep the engine speed constant during the cruise and if this is done the thrust remains nearly constant. If both the thrust and the specific fuel consumption c are constant the mass of fuel used per unit time is also constant and the specific range is directly proportional to V. (This is also shown by Equation (7.1), since $F = D = \beta W$, so that βW is constant.) Hence in Figure 7.3 a cruise at constant thrust is represented by a straight line passing through the origin, such as the line EF which is shown. An increase in the value of the constant thrust would rotate the line EF about the origin, shifting it to the right, corresponding to greater v and β, but smaller r_a for a given speed. This procedure is used only for relatively short ranges because it does not compare well over long ranges with either of the alternatives represented by AB and CD. In comparison with AB, the line EF gives the same order of speed variation but a lower mean specific range, because v is below the optimum value in the early part of the cruise and above it in the later part. In comparison with CD, the line EF gives a mean specific range that is usually lower and a mean speed that is always lower, if the height is chosen so that the speeds at the points D and F both give Mach numbers close to M_D.

In a cruise either at constant v or at constant thrust, represented either by the line AB or by EF in Figure 7.3, there is a small acceleration (negative for AB) but in a typical case this is only of order 5 m/s per hour and the effective value of β is changed by only about 0.25%.

7.6 Conditions for maximum cruise range – propellers

For a propeller aircraft the maximum speed for economical cruising is usually limited by the Mach number M_P discussed in § 5.3, above which the effects of high Mach number on the propeller cause unacceptable noise and some loss of efficiency. For the older conventional forms of propeller M_P is about 0·6 but for modern advanced propellers and propfans it may be as high as 0.8.

182 *Fuel consumption, range and endurance*

The specific fuel consumption of a propeller aircraft is usually defined as $c' = Q/P_e$, where Q is the mass of fuel used per unit time and P_e is the equivalent shaft power defined by Equation (5.17). (For a piston engine the exhaust jet thrust is negligible and P_e is simply the shaft power.) The specific range is given by Equation (7.6) and for any given weight this is maximum when $c'\beta/\eta_{PR}$ is minimum. For conditions close to maximum cruise power both η_{PR} and c' may often be assumed to be independent of speed and height, although for a given height there is a slight decrease of c'/η_{PR} with increasing speed as given by Equation (5.24), and there may also be some reduction of c' with increasing height as indicated in Figures 5.20 and 5.21. When the power is below the maximum cruise level there is usually a further reduction of c' as the height is increased, as explained in §5.4.2, because of the rise in the ratio P_s/P_{sm} as the rated power P_{sm} decreases.

If η_{PR} and c' are assumed to be independent of speed, for any given height, the condition for maximum specific range is that β should have the minimum value β_m and to maintain this condition throughout the cruise C_L is required to remain constant and equal to C_L^*. As for turbofan aircraft this can be achieved by a cruise-climb at a constant true air speed V with $\sigma \propto W$ and the total cruise range is then given by Equation (7.8).

With $\beta = \beta_m$ and $v = 1$ an increase in the height at which the cruise-climb is started gives an increase in the constant cruise speed V and usually also an increase of range because of the effect of height on c' when the power is below the maximum cruise level. Thus there is a double benefit from increasing height, as there is for turbofans, and if sufficient power is available it is best to choose the height at the start of the cruise-climb so that the greatest Mach number reached in the cruise is close to the value M_P mentioned earlier. Since V and β are to remain constant during the cruise-climb the required thrust power is

$$DV \propto D \propto W \propto \sigma$$

and for $h > 11$ km in the ISA the shaft power with constant engine setting is proportional to σ, so that no adjustment of the engine setting should be needed during the cruise. More often with this class of aircraft the whole of the cruise is completed at heights below 11 km and the variation of temperature with height then makes it necessary to make small adjustments of engine setting during the cruise.

Figure 7.4 is a diagram for a propeller aircraft cruising at constant height, similar to the turbofan diagram of Figure 7.3. The height of 8 km is the same for the two cases and the two aircraft represented have the same wing area, drag characteristics and initial and final masses. For the propeller aircraft of Figure 7.4 it is assumed that η_{PR} and c' have the constant values of 0.85 and 7×10^{-8} kg/J respectively.

Conditions for maximum cruise range – propellers

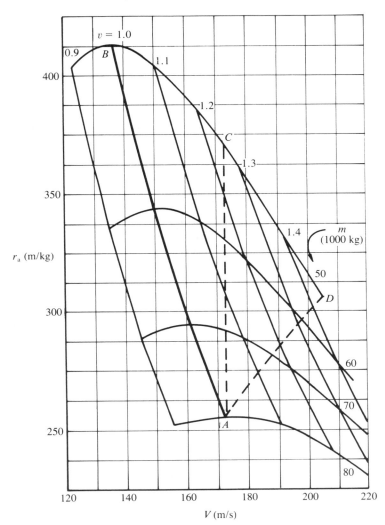

Figure 7.4. Variation of specific range with speed and aircraft mass at constant height – propeller aircraft. $c' = 7 \times 10^{-8}$ kg/J, $\eta_{PR} = 0.85$, $h = 8$ km, $S = 150$ m^2, $K_1 = 0.02$, $K_2 = 0.045$.

Since turboprop aircraft are used mainly over short ranges the large ratio of initial to final mass represented in Figure 7.4 is somewhat unrealistic, but it has been chosen in order to illustrate the effects of varying mass on the various cruise procedures.

The line AB in Figure 7.4 represents the propeller aircraft cruising at constant height, with the optimum condition $\beta = \beta_m$ maintained by reducing the speed progressively so that $V \propto W^{1/2}$ and C_L remains constant. The disadvantage of this procedure is similar to that of the cruise shown by the line AB in Figure 7.3, namely that the Mach number must not be greater than M_P at the start and unless the range is short the speed in the later part of the flight will be substantially lower, giving an unacceptably low average speed. A further objection to this procedure is that the required thrust power DV falls substantially during the cruise because both D and V decrease, so that frequent adjustments of the engine controls are needed and moreover the reduction of power may lead to an increase of c'.

The two other cruise procedures to be considered at constant height are represented by the lines AC and AD in Figure 7.4. In AC the speed is kept constant and the average speed can be high, but the speed ratio v increases during the cruise and β rises above the minimum value, so that the mean specific range is less than that obtained with AB. In Figure 7.4 the line AC is shown as starting at $v = 1$ but this is not essential; a start at a lower value of v would give a greater range, although there are objections to operation at v less than 1 which will be discussed in §7.7. A reduction of height would decrease the speeds for all the cruise procedures shown in Figure 7.4 and would also reduce the range, because the required power would become a smaller proportion of the maximum available power and this would increase c'. The effect of height on specific range is shown by some results based on a survey of turboprop aircraft given in ESDU 75018 (1975). These show that the maximum specific range increases with height by an amount between 8 and 13% per km near sea level and 3–8% per km at a height of about 7.5 km. Data given by Dornier GmbH (1986) for the Dornier 228 aircraft are consistent with these figures and show an increase of maximum specific range with height of about 9% per km for heights between 2 and 5 km. Since increasing height is clearly beneficial, the best procedure if sufficient power is available is to choose the height so that the Mach number is equal to M_P with the selected value of v at the start.

In the constant speed cruise of AC the drag decreases as the weight falls and hence the required thrust power decreases, although not as much as for AB. Thus adjustments of the engine controls are needed during the cruise and there may also be an increase of c' as the power is reduced. To avoid this a cruise at constant power is sometimes used; then if the propeller efficiency is constant DV is constant and the conditions are represented by a line such as AD in Figure 7.4. As for the line AC, the starting point is assumed here to be at $v = 1$ but this is not essential. The constant power procedure is unsatisfactory unless the range is short, for two reasons which can be seen from Figure 7.4. First, the mean specific range is low because v rises well above 1 in the later part of the cruise and the point D is well removed from the maximum on the curve for the lowest m. (This could be avoided only by making v too low at the start to be practicable.) Second, the Mach number must not rise above M_P at the end of the cruise and unless the range is short the speed at the start and the average speed will be unacceptably low. It should be noted that for a short-range flight the mass of fuel used and hence the change of mass would be small and thus the point A would be only a little below B. In this case the different cruise procedures would give closely similar ranges and a cruise at constant speed would normally be used.

The results of the survey of turboprop aircraft given in ESDU 75018 (1975) show that in practice the value of v for maximum r_a is usually found to be greater than 1 and decreases with increasing height from values between 1.2 and 1.5 near sea level to values between 1.0 and 1.2 at 7 or 8 km height. The difference between these values and the theoretical result $v = 1$ can be explained only by variations of η_{PR} and c' with speed. Near sea level the true air speed is relatively low for values of v near 1 and there may often be an increase of η_{PR} with speed as indicated in Figure 5.18. This would increase the value of v for minimum $c'\beta/\eta_{PR}$ and for maximum r_a. At higher altitudes the true air speed is greater and the variation of η_{PR} with speed is likely to be smaller. The main reason for the optimum value of v being greater than 1 may then be a reduction of c' with increasing speed, although the small reduction given by Equation (5.25) changes the theoretical value of v for maximum r_a only from 1.0 to 1.025.

ESDU 75018 (1975) also gives results from a survey of range data for piston-engine aircraft. These show that the value of v for maximum specific range is typically between 1.05 and 1.2, sometimes decreasing slightly with increasing height. The survey also shows some increase of maximum specific range with height but this is less than that found in the survey of turboprops, being typically about $1\frac{1}{2}\%$ per km at 3–5 km height.

7.7 Practical cruise procedures

For an aircraft powered by turbofan engines, with c increasing only gradually with speed in accordance with Equation (5.11), the specific range as given in Equation (7.1) increases with true air speed if β is near to the minimum value. It is therefore usual to fix the cruising Mach number of the aircraft at a value close to M_D as explained in § 7.5.1. If a cruise-climb over any desired range of heights is allowed by ATC and if ample thrust is available, maximum range is obtained theoretically in a cruise-climb with $v = 1$ throughout, but it is usual to fly at a value of v somewhat greater than 1, say 1.15. This gives a substantial reduction of cruise lift coefficient (since $C_L \propto v^{-2}$), with only a small increase of β (e.g. $\beta/\beta_m = 1.02$ for $v = 1.1$), and the reduction of lift coefficient is valuable in keeping the cruising condition farther away from the *buffet boundary*. This is the limiting value of C_L above which flow separation on the upper surface of the wing leads to wing buffet and sometimes to other undesirable effects. The separation is caused by the shock wave which forms on the upper surface of the wing, at the downstream boundary of the region of locally supersonic flow which develops as the flight Mach number reaches and exceeds M_D. Any increase of either M or C_L increases the strength of this shock wave, so that the value of C_L at the buffet boundary decreases as M rises. The normal cruising condition must always be clear of the buffet boundary but there is often a real possibility of buffeting in severe atmospheric turbulence, because of the short-term increase in the angle of incidence induced by an upward gust, and a reduction of cruise lift coefficient is beneficial in avoiding this.

As explained in § 7.5.4 a reduction in the installed thrust, below the value required for cruising at $M = M_D$ and $v = 1$, can give a useful reduction of engine weight with only a small reduction of specific range. This is the basis of the usual optimum design for a turbofan aircraft, with v in the region of 1.15 as given by Equation (7.14) or (7.15).

As mentioned earlier a stepped cruise may sometimes be allowed by ATC as an alternative to a cruise-climb. If there is a free choice of heights for the steps it is usual to keep M constant, close to M_D, and choose the heights so that at the *start* of each level phase v is in the region of 1.15, thus ensuring that it never falls below this value. More often, and especially for short to medium ranges, the aircraft is required to fly at a constant height for the whole of the cruise. If there is a choice of this height the usual procedure is the same as for each level phase of the stepped cruise, keeping M constant and close to M_D and $v \approx 1.15$ at the start. In some cases the constant height is specified by ATC and the theoretical optimum value of v is then in the region of 1.15 to 1.2 as given by Equation (7.18). In order to keep the flight time as short as possible it is usual also in this case to keep M close to M_D, even if this means that v is above 1.15 at the start and well above the optimum for most of the cruise distance, although of course there will then be a drag penalty and the fuel consumption will be increased.

For a turboprop aircraft the Mach number for economical cruising is limited to the value M_P explained in § 5.3, above which there is a high noise level and some loss of efficiency due to effects of high Mach number at the blade tips. For aircraft in service at present (1990) M_P is only about 0.6 and for this reason turboprop aircraft are normally used mainly for short ranges, and not for the longer ranges over which passengers expect and obtain greater speeds. An example of a modern

Practical cruise procedures 187

Figure 7.5. The SAAB 2000 regional turboprop aircraft (shown here as a scale model).

turboprop aircraft is the SAAB 2000 shown in Figure 7.5, which has Dowty Rotol propellers with sweptback blades to maintain high efficiency and reduce the noise in the cabin, even though the normal cruising Mach number is only about 0.6. With further development of propfans and other advanced forms of open rotor in the future, M_P may be raised to 0.8 or more and aircraft using these propfans are then likely to be used for both long and short ranges, with cruising speeds and heights determined by arguments similar to those used for turbofan aircraft.

For the short-range turboprop aircraft in service at present both the speed and the height are normally kept constant during the cruise and the Mach number is chosen to be close to M_P to minimise the time of flight. The weight of a turboprop power plant, with its propeller and reduction gear, is usually greater than that of a turbofan that gives the same cruising thrust and in many cases the weight is minimised by installing only sufficient power to satisfy the requirements of take-off and initial climb after failure of one engine. The cruising height is then chosen as the greatest at which the Mach number M_P can be maintained at the start of the cruise with the power that is available, because this ensures that maximum cruise power is used at the start, so that c' is as low as possible, and also exploits the fact that increasing height allows M_P to be reached at a lower value of v, closer to the optimum. For short range there is only a small increase of v as the weight of the aircraft falls during the cruise and with the speed and height determined in this way the mean value of v may be about 1.3 or 1.4 in a typical case. Greater installed power would allow greater height and hence a lower value of v, but this would increase the weight of the engine and propeller and in many cases would increase the combined weight of fuel and power plant, thus reducing the payload.

It has already been mentioned that there is a beneficial effect of increasing v because C_L is then reduced and thus the cruising condition is moved farther away from the buffet boundary. This is more important for turbofans than for the turboprops in service at present because of the higher cruising Mach numbers of the former, giving lower values of C_L at the buffet boundary. A further reason for increasing v above 1, and for avoiding flight with $v<1$, is the desirability of achieving adequate speed stability as discussed in § 2.7. For both these reasons and because of the need to keep the engine weight low, especially for short ranges where fuel weight may be less important than engine weight, the values of v used for cruising are usually above 1.1 and may be as high as 1.5, especially for propeller aircraft operating over short ranges.

7.8 Calculation of cruise range

For aircraft powered either by turbofans or by turboprops the theoretically ideal procedure giving maximum cruise range is a cruise-climb at constant speed or Mach number, with v kept at the optimum value which is either 1 or close to 1, as discussed in § 7.5.1, 7.5.2 and 7.6. The range calculated for this form of cruise is often known either as the *Breguet range*, as mentioned earlier, or the *reference range* and will be denoted here by the symbol R_{acr}. This reference range is easily calculated from Equation (7.3) or (7.8) and is useful as a standard of comparison, although the cruise procedure giving $R_{ac} = R_{acr}$ is not likely to be used in practice. Using values of cruise range R_{ac} calculated for various cruise procedures, the corresponding range ratios (R_{ac}/R_{acr}) can be found and these give a useful measure of the loss of range caused by departure from the theoretically ideal procedure. Analytical expressions for range ratios are given in ESDU 73019 (1982) for various cruise procedures with the same *initial* conditions as those used in calculating R_{acr}. Peckham (1970) also gives expressions of this kind and points out that it may sometimes be more realistic to make comparisons with a second case which assumes the same initial weight and height but for which the initial cruise speed is that which gives maximum range, rather than the initial speed used in calculating R_{acr}. In any case, in giving either an analytical expression or a numerical value for the range ratio it is necessary to specify the conditions assumed in calculating both R_{ac} and R_{acr}. (In calculating R_{acr}, either V or M may be assumed to be constant and the value chosen must be specified.)

Calculation of cruise range

In the preceding sections of this chapter the emphasis has been on discussion of the value of the speed ratio v to be used in the cruise, for various specified conditions such as constant speed or height. When v is known, the values of all the quantities required for the calculation of range can be found from the specified cruise conditions, the drag characteristics of the aircraft and the specific fuel consumption and this will be illustrated in the following examples. Analytical expressions for range ratios are not derived here because numerical calculation of range is simple and straightforward, using either Equation (7.3) or (7.8) or, more generally, a numerical evaluation of the integral

$$R_{ac} = -\int_{m_1}^{m_2} r_a \, dm,$$

as explained earlier in discussion of Equation (7.2).

EXAMPLE 7.3. RANGE OF TURBOFAN AIRCRAFT, WITH ALTERNATIVE CRUISE PROCEDURES

A turbofan aircraft has a wing area of 160 m² and a drag coefficient given by Equation (3.6) with $K_1 = 0.02$ and $K_2 = 0.045$. The specific fuel consumption of the engines is given by

$$c = c_1 \theta^{1/2} M^n \quad \text{with} \quad c_1 = 2.2 \times 10^{-5} \text{ kg/N s}$$

and

$$n = 0.45.$$

The mass at the start of the cruise is 100 000 kg and 30 000 kg of fuel is available for use in the cruise.

I. Find the cruise range for the following conditions (a) to (d), all at a constant true air speed of 236 m/s:

  (a) A cruise-climb with $v = 1$.
  (b) A cruise-climb with $v = 1.15$.
  (c) A cruise at a constant height, chosen to give $v = 1.15$ at the start.
  (d) A stepped cruise with the height constant at the value used in (c) until half the fuel is used. There is then a climb to a level giving $v = 1.15$ and this height is held constant for the remainder of the cruise.

II. Find the maximum cruise range if the aircraft is required to fly at a constant height of 9 km, with the true air speed kept constant and with v never less than 1.15.

Preliminary calculations from Equations (3.13) and (3.17) lead to

$$\beta_m = 2(K_1 K_2)^{1/2} = 0.06$$

and

$$V_e^* = (2w/\rho_0)^{1/2}(K_2/K_1)^{1/4} = 1.565 w^{1/2} \text{ m/s}.$$

At the start, the given mass and wing area show that

$$w = 6131 \text{ N/m}^2 \quad \text{and} \quad V_e^* = 122.54 \text{ m/s}.$$

I(a). Cruise-climb with $v = 1$.

At the start, $\sigma = (V_e^*/V)^2 = 0.2696$, which requires the height to be 11.63 km. Therefore the whole of the cruise will be at $h > 11$ km, for which $\theta^{1/2} = a/a_0 = 0.8671$, $a = 295.1$ m/s and thus $M = 0.800$, leading to a specific fuel consumption of $c = 1.725 \times 10^{-5}$ kg/N s. The optimum value $v = 1$ allows β to have its minimum value $\beta_m = 0.06$ and since V, c and β are all constant the cruise range is given by Equation (7.3) as

$$R_{ac} = 23\,244 \ln(100/70) \text{ km} = 8290 \text{ km}$$

and this is the reference range R_{acr}.

I(b). Cruise-climb with $v = 1.15$.

With this value of v, Equation (3.18) gives $\beta = 0.0624$ and the initial value of the relative density is

$$\sigma = (vV_e^*/V)^2 = 0.3566$$

for which the appropriate height is 9.56 km, now less than the critical value of 11 km. This initial height leads to

$$\theta^{1/2} = a/a_0 = 0.8859, \quad M = 0.783$$

and

$$c = 1.746 \times 10^{-5} \text{ kg/N s},$$

but as the flight will ultimately rise above 11 km, the conditions at $h = 11$ km are also required. These are $c = 1.725 \times 10^{-5}$ kg/N s as in I(a), $\sigma = 0.2971$ and, since σ must change in proportion to the aircraft mass, $m = (0.2971/0.3566) \times 10^5$ kg $= 0.8331 \times 10^5$ kg. Thus the mass falls from 10^5 to 0.8331×10^5 kg as the aircraft climbs from 9.56 to 11 km. During this phase of the flight c varies slightly, with a mean value of 1.736×10^{-5} kg/N s. Using this mean value, Equation (7.3) gives the range for this phase as

$$22\,208 \ln(1/0.8331) \text{ km} = 4055 \text{ km},$$

whereas for the remainder of the cruise $c = 1.725 \times 10^{-5}$ kg/N s as in I(a) and the range is

$$22\,350 \ln(0.8331/0.7) = 3890 \text{ km}.$$

Thus the total range is 7945 km and the range ratio is $R_{ac}/R_{acr} = 0.958$.

As a guide to the saving of engine weight in changing the cruise procedure from (a) to (b) it is of interest to compare the values of F/σ, because for $h \geq 11$ km this quantity is roughly proportional to engine weight. Since $F/\sigma = \beta W/\sigma$ and $W \propto \sigma$ during the cruise-climb,

Calculation of cruise range

the comparison can be made for any selected value of W, e.g. choosing the conditions at the start,

$$\frac{(F/\sigma)_b}{(F/\sigma)_a} = \frac{(\beta/\sigma)_b}{(\beta/\sigma)_a} = \frac{0.0624}{0.3566} \times \frac{0.2696}{0.06} = 0.786.$$

Thus a change of v from 1.0 to 1.15, to allow optimum performance at lower altitudes when thrust is a limitation, while departing from the minimum drag condition of $v = 1$, gives a reduction of F/σ of 21%, with a reduction of range of only about 4%. This shows how the engine weight can be reduced substantially, with only a small loss of range.

I(c). Cruise at constant height with $v = 1.15$ at the start.

The values of σ, h, θ, M and c are constant throughout the cruise and are the same as for case (b) at the start. For fixed values of σ and V, but allowing for changes in the mass m,

$$v \propto V_e^{*-1} \propto m^{-1/2}.$$

Thus $v = 1.15(10^5/m)^{1/2}$ and then $\beta = 0.03(v^2 + v^{-2})$. Using these results the specific range r_a can be found from Equation (7.1) for any given mass. If the total cruise is divided into 6 stages, with 5000 kg of fuel used in each, the mean value of r_a can be found for each stage and the total cruise range is then

$$R_{ac} = 5 \sum r_a \text{ km, with } r_a \text{ in m/kg.}$$

This gives $R_{ac} = 7378$ km and the range ratio R_{ac}/R_{acr} is 0.890. Thus the change from a cruise-climb to a cruise at constant height, with $v = 1.15$ at the start in both cases, reduces the range by about 7%.

I(d). Stepped cruise with $v = 1.15$ at the start of each level phase.

For the first level phase, while the mass falls from 10^5 to 0.85×10^5 kg, all the conditions are the same as for I(c) and the range for this phase is given as 3500 km by the first three stages of the calculation of the range in case I(c). For the second level phase the mass at the start is 85 000 kg so that the density ratio required to give $v = 1.15$ at the start has dropped to

$$\sigma = (m_2/m_1) \times \sigma_1$$
$$= 0.85 \times 0.3566 = 0.3031.$$

Hence the values appropriate to this second phase are

$$h = 10.84 \text{ km}, \qquad \theta^{1/2} = 0.8691, \qquad M = 0.798$$

and

$$c = 1.727 \times 10^{-5} \text{ kg/N s.}$$

When the mass is m, $v = 1.15(85\,000/m)^{1/2}$ and as before $\beta = 0.03(v^2 + v^{-2})$. Hence r_a can be found for any given mass. Dividing this phase of the cruise into 3 stages, using 5000 kg of fuel in each, the range for this phase is found to be 4200 km, giving a total cruise range of 7700 km. The corresponding range ratio is 0.929 and the introduction of the step to a greater height has increased the range by 4.4% in comparison with the constant height cruise of I(c).

No allowance has been made in this calculation for the increased amount of fuel used in the step climb from the first level phase to the second. The height gained in the step is 1.28 km and if the corresponding gain of potential energy is equated to the thrust work required just to fly horizontally for a distance Δs km, there will be the equivalence

$$\beta W \Delta s = 1.28 W,$$

and with the mean value of β equal to 0.064 this gives $\Delta s = 20$ km. This means that the extra fuel needed for the gain of potential energy would be sufficient, if there were no climb, for a further 20 km of level cruise. Thus the range calculated for the stepped cruise should strictly be reduced by 20 km, but the correction is of no importance, being only about 0.25% of the full range.

II. Cruise at a specified constant height.

For this case Equation (7.18) shows that for $n = 0.45$ the optimum value of v is 1.152. In a cruise with both height and speed constant, v increases as the weight falls, so that if v is required to be never less than 1.15 the optimum conditions are obtained by starting the cruise at $v = 1.15$. For the specified height of 9 km, $\sigma = 0.3807$, $\theta^{1/2} = 0.8927$ and the constant true speed is found from the starting conditions as

$$V = 1.15 V_e^* \sigma^{-1/2} = 228.4 \text{ m/s}.$$

Hence $M = 0.752$ and $c = 1.728 \times 10^{-5}$ kg/N s. At any mass m the values of v and β are the same as for case I(c) and since $r_a \propto V/c$, for given β and m, the ratio of r_a for this case to that of case I(c) is

$$\frac{228.4}{1.728} \times \frac{1.746}{236} = 0.978.$$

Thus the range R_{ac} for this case is $0.978 \times 7378 = 7216$ km, giving a range ratio of 0.870. The reduction of range in comparison with case I(c) is due to the reduced height, giving lower true air speed, and if the specified height had been still lower the reduction of range would have been even greater.

EXAMPLE 7.4. RANGE OF TURBOPROP AIRCRAFT, WITH ALTERNATIVE CRUISE PROCEDURES

A turboprop aircraft has a wing area of 100 m² and a drag coefficient given by $K_1 = 0.02$ and $K_2 = 0.045$. The specific fuel consumption c' of

Calculation of cruise range

the turboprop engines is 6×10^{-8} kg/J and the propeller efficiency η_{PR} is 0.85. Both c' and η_{PR} may be assumed to be constant within the range of speeds and heights considered for cruising. The mass at the start of the cruise is 50 000 kg and 10 000 kg of fuel is available for use in the cruise. Find the cruise range for the following conditions:

(a) A cruise-climb at a constant true air speed of 180 m/s, with $v = 1$.
(b) A cruise at constant height and speed, with $V = 190$ m/s and $v = 1.15$ at the start.
(c) As (b), but with $v = 1.3$ at the start.

As in Example 7.3, preliminary calculations lead to $\beta_m = 0.06$ and $V_e^* = 1.565 w^{1/2}$, but in this case, at the start, $w = 4905$ N/m² and $V_e^* = 109.6$ m/s.

(a) Cruise-climb with $v = 1$.
 At the start, $\sigma = (V_e^*/V)^2 = 0.3707$, for which the height h is 9.22 km.
 At the end, with $\sigma \propto m$, $\sigma = (4/5) \times 0.3707 = 0.2966$, which implies a rise to $h = 11.01$ km. (The greatest Mach number is reached at the end of the cruise and is 0.610.) Throughout the cruise $\beta = \beta_m = 0.06$ and the range is given by Equation (7.8) as

$$R_{ac} = 24\,068 \ln(5/4) = 5371 \text{ km}$$

and this is the reference range R_{acr}.

(b) Cruise at constant height and speed, with $V = 190$ m/s and $v = 1.15$ at the start.
 The conditions at the start show that

$$\sigma = (1.15 \times 109.6/190)^2 = 0.4401$$

and

$$h = 7.77 \text{ km}.$$

(Although V is higher than in (a) the height is lower and the Mach number is only 0.615, about the same as the maximum value in (a).) At any mass m, $v = 1.15(50\,000/m)^{1/2}$ and the specific range is given by Equation (7.6) as

$$r_a = \eta_{PR}/(c'\beta \, mg) = (1.444 \times 10^6)/(\beta m) \text{ m/kg}.$$

Dividing the total cruise into four stages, with 2500 kg of fuel used in each stage, the mean value of r_a can be found for each stage and the total cruise range is then

$$R_{ac} = 2.5 \sum r_a \text{ km} \quad \text{(with } r_a \text{ in m/kg)}.$$

This gives $R_{ac} = 4977$ km $= 0.927 R_{acr}$.

(c) Cruise at constant height and speed, with $V = 190$ m/s and $v = 1.3$ at the start.

In this case the conditions at the start show that

$$\sigma = (1.3 \times 109.6/190)^2 = 0.5623$$

and

$$h = 5.62 \text{ km}.$$

The Mach number is 0.597, $v = 1.3(50\,000/m)^{1/2}$ and the equation giving r_a is the same as that used for case (b). Dividing the cruise into four stages as in (b), the range is found to be

$$R_{ac} = 4437 \text{ km} = 0.826 R_{acr}.$$

The range in this case is considerably less than for (b) because of the higher values of v and consequent higher β. The power required is maximum at the start of the cruise and is higher for (c) than for (b) in the ratio in which β is increased. This power ratio is obtained by using Equation (3.18) as

$$(1.3^2 + 1.3^{-2})/(1.15^2 + 1.15^{-2}) = 1.098,$$

but for (c) the maximum power is required at a much lower height (5.62 instead of 7.77 km) so that the weight of engine required for (c) is likely to be less than for (b), indicating that for an aircraft of short to medium range the best choice of engine may be one of modest thrust, provided this meets the requirements of take-off and climb. The cruising height may then be fairly low and the speed ratio v may be relatively large.

7.9 Endurance

As mentioned earlier, endurance is usually much less important than range but is valuable for some special purposes such as ocean patrol. The requirement is then that the aircraft should remain in the air for the longest possible time for a given consumption of fuel and the distance covered is usually of secondary importance. For these special purposes the aircraft usually flies at a constant height and sometimes at a constant speed, although in a long flight the speed will probably be adjusted as the weight falls to keep the conditions close to the optimum for long endurance.

Endurance is also important for an aircraft holding in a stack, awaiting permission to land. The height is usually specified by ATC but there is often a choice of speed so that the flight conditions can be close to those required for maximum endurance, giving minimum consumption of fuel for a given time in the stack.

The *specific endurance* is defined as the time flown per unit mass of fuel used and is

$$e_t = 1/Q = -\mathrm{d}t/\mathrm{d}m = r_a/V, \tag{7.19}$$

Endurance

where r_a is the specific range. The *endurance* is the total time of flight and is

$$E_t = -\int_{m_1}^{m_2} e_t \, dm. \tag{7.20}$$

This is maximum if e_t can be maximised at every instant of the flight.

7.9.1 TURBOFANS

For a turbofan aircraft, Equations (7.1) and (7.19) show that

$$e_t = 1/(c\beta W) \tag{7.21}$$

and if c is constant this is maximum for a given weight when $v = 1$, so that $\beta = \beta_m$. For a given height it may sometimes be acceptable to assume that c is constant but it is more accurate to represent the variation of c with speed by Equation (5.11), giving $c \propto M^n$. The optimum value of v is then the value giving minimum βv^n and this is $v = [(2-n)/(2+n)]^{1/4}$. For positive values of n this is less than 1, e.g. for $n = 0.5$ the optimum v is 0.88, but operation with low values of v raises the question of speed stability.

If c and β are constant the endurance is given by Equations (7.20) and (7.21) as

$$E_t = 1/(c\beta g) \int_{m_2}^{m_1} dm/m = [1/(c\beta g)] \ln(m_1/m_2) \tag{7.22}$$

and this is maximum when $\beta = \beta_m$.

When there is a free choice of height the best choice depends on the way in which c varies with height, for constant v, i.e. constant EAS. Decreasing height has the following three effects, the first two being explained by Equation (5.11):

(a) A decrease of Mach number, tending to reduce c.
(b) A rise of temperature, tending to increase c.
(c) An increase in the available thrust F_m at the cruise rating.

With no change in the required thrust, for given β and W, because $D = \beta W$, the effect (c) causes a decrease of F/F_m and, as illustrated in Figure 5.12, this may decrease c if the change of F/F_m is small, although for a large change there is always an increase of c. This effect (c) is likely to be dominant in many cases and there will then be a decrease of c for a small reduction of height but an increase for a large loss of height. Use of Figure 5.12 as a guide suggests that for minimum c the height should be chosen so that the required thrust is roughly 20% below the rated cruise value.

In practice, when long endurance is required the value of v at the start is usually chosen to be above the theoretical optimum in order to ensure positive speed stability. A value in the region of 1.1 or 1.2 is commonly used and if this is maintained by reducing the speed as the weight falls, at constant height, or by increasing the height at constant speed, the endurance obtained is only slightly lower than the maximum. If both the height and the speed are kept constant v increases as the weight falls and this causes a loss of endurance. When there are four engines the endurance can sometimes be increased by shutting down two of the engines, so that the remaining engines operate at a higher value of F/F_m and hence with a lower specific fuel consumption.

EXAMPLE 7.5. ENDURANCE OF TURBOFAN AIRCRAFT, WITH ALTERNATIVE FLIGHT PROCEDURES

The aircraft described in Example 7.3 may be assumed to have a constant specific fuel consumption $c = 1.6 \times 10^{-5}$ kg/N s. Find the endurance for flight at a constant height, with a starting mass of 100 000 kg and 30 000 kg of fuel available for use, for the following conditions:

(a) $v = 1$ throughout.
(b) $v = 1.15$ throughout.
(c) Constant true air speed with $v = 1.15$ at the start.

(a) $v = 1$ throughout.
The speed must fall gradually to keep $V \propto W^{1/2}$. With $\beta = \beta_m = 0.06$, Equation (7.22) gives

$$E_t = (1.062 \times 10^5) \ln(10/7) \text{ seconds} = 10.52 \text{ hours}.$$

(b) $v = 1.15$ throughout.
As for (a), the speed must fall gradually. $\beta = 0.06236$ and Equation (7.22) shows that E_t is reduced in the ratio β_m/β. Thus

$$E_t = 10.12 \text{ hours}.$$

(c) Constant true speed with $v = 1.15$ at the start.
As for case I(c) of Example 7.3, $v = 1.15(10^5/m)^{1/2}$, $\beta = 0.03(v^2 + v^{-2})$ and, using Equation (7.21) with $W = mg$, the endurance given by Equation (7.20) is

$$E_t = \frac{1}{cg}\int_{m_2}^{m_1} \frac{dm}{\beta m}, \quad \text{with } 1/(cg) = 6371 \text{ seconds}.$$

Dividing the total flight into 6 stages, as in case I(c) of Example 7.3, with 5000 kg of fuel used in each stage, the endurance is found to be

$$6371 \times 5000 \sum [1/(\beta m)] \text{ seconds} = 9.47 \text{ hours}.$$

7.9.2 TURBOPROPS

For a turboprop aircraft, Equation (7.6) shows that

$$e_t = r_a/V = \eta_{PR}/(c'\beta WV) \tag{7.23}$$

and for a given weight this is maximum when $c'\beta V/\eta_{PR}$ is minimum. If it can be assumed that c'/η_{PR} is constant, the optimum condition is that βV should be minimum and this is the condition for minimum drag power. As shown in § 3.4, $\beta V \propto C_D/C_L^{3/2}$ for a given height and if the simple parabolic drag law is valid this is minimum when $v = 3^{-1/4} = 0.76$, but attention is drawn to the remarks in Chapters 2 and 3 about the closeness of the corresponding speed to the stalling speed and the possible invalidity of the simple parabolic drag law.

In reality c'/η_{PR} is not exactly constant, even at a fixed height. When the height is large enough for the power to be close to the maximum cruise level the variation of c'/η_{PR} with speed is given with fair accuracy by Equation (5.24), with $n = 0.9$. Then for a fixed height the condition for maximum e_t is that $\beta v^{0.9}$ should be minimum and the required speed ratio is

$$v = (1.1/2.9)^{1/4} = 0.785. \tag{7.24}$$

At lower heights the required power is a smaller proportion of the maximum cruise power and the validity of Equation (5.24) is doubtful. Moreover the true air speed is lower and as a result η_{PR} may increase with speed as indicated in Figure 5.18 for low values of J. Thus the decrease of c'/η_{PR} with increasing speed may be more rapid than that given by Equation (5.24) and hence the optimum value of v may be greater than 0.785.

As the height is reduced at a constant value of v, i.e. at a constant EAS, there is a reduction of true air speed and for the reasons already explained this leads to an increase of c'/η_{PR}, especially at low heights. In addition there is a significant effect of reduced power, as explained in § 5.4.2 with reference to Figure 5.22. With decreasing height the required thrust power DV is reduced in proportion to V, because there is no change of drag, yet the available maximum cruise power increases. Hence there is a substantial reduction in the ratio P_s/P_{sm} shown in Figure 5.22 and a consequent increase of c'. Thus the maximum specific endurance usually increases with height and an example of this is shown by the curve for 2 engines in Figure 7.6, which has been derived from data given by Dornier GmbH (1986) for the Dornier 228 aircraft. Comparison of the two curves in Figure 7.6 shows that for this aircraft the maximum specific endurance is increased by about 50% when one of the engines is shut down. The true air speed for maximum e_t is slightly reduced when only one engine is used, so that any change of η_{PR} would be a reduction, showing that the large gain of e_t in this case can only be caused by a reduction of c' due to the increase of P_s/P_{sm}. The shutting down of one engine would probably not be used in practice as a means of increasing the endurance of a twin-engine aircraft such as the Dornier 228, because of undesirable effects of the asymmetry of flight, but on a four-engine aircraft two of the engines are sometimes shut down when long endurance is required, just as they may be for turbofan aircraft.

In choosing a practicable cruising speed for a flight in which long endurance is required an important consideration is the need to ensure adequate speed stability. This means that the EAS should be well above the speed V_{ec} discussed in Chapter 2 and shown in Figure 2.9. The f curves shown in Figure 2.9 for different heights can be regarded equally well as curves for a constant height at different power settings. The f curve that just touches the β curve corresponds to the lowest power setting at which horizontal flight can be maintained and at the point where the curves touch the EAS V_e is equal to V_{ec}. Thus if FV is assumed to be constant for any given power setting, V_{ec} is the EAS for minimum drag power and if the simple parabolic drag law is valid $v = 0.76$ at this condition. In reality FV is not exactly constant and moreover the value of v for minimum drag power may be somewhat greater than 0.76, because the aircraft may be approaching the stalling condition and the simple parabolic drag law may not be valid, but V_{ec} is not likely to be greater than about $0.85\,V_e^*$ in most cases. Thus it may be assumed that in order to avoid speed instability the speed ratio

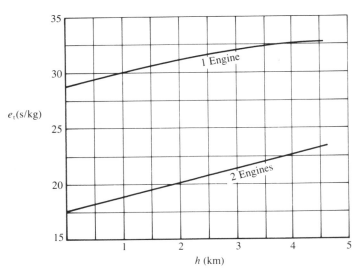

Figure 7.6. Maximum specific endurance of Dornier 228 ($m = 5000$ kg) from data given by Dornier GmbH (1986).

v should be greater than about 0.85, or perhaps 0.9. ESDU 75018 (1975) reports that in order to ensure adequate speed stability the values of v used for 'holding' propeller aircraft are in the range 0.9–1.05, while for flights of long duration where long endurance is required, v is usually between 1.0 and 1.2.

EXAMPLE 7.6. ENDURANCE OF TURBOPROP AIRCRAFT, WITH ALTERNATIVE FLIGHT PROCEDURES

For the aircraft described in Example 7.4, find the endurance for flight in the following conditions, with a starting mass of 50 000 kg and 10 000 kg of fuel available for use, assuming that η_{PR} and c' have the constant values 0.80 and 8.5×10^{-8} kg/J respectively. (Because of the reduced power and low flight speed, c' is greater than it was in Example 7.4 and η_{PR} is slightly lower.) The conditions are:

(a) The height is kept constant at 2 km, with the speed continually adjusted to keep $v = 0.76$.
(b) As (a) but with $v = 1.0$.
(c) The height is constant at 2 km and a constant speed is chosen to give $v = 1.0$ at the start.

As in Example 7.4, $\beta_m = 0.06$ and at the start $V_e^* = 109.6$ m/s. If the mass is m kg at any point after the start,

$$V_e^* = 109.6 \, (m/50\,000)^{1/2} \text{ m/s.}$$

(a) For $h = 2$ km and $v = 0.76$, the necessary constants are

$$\beta = 0.0693, \quad V = vV_e^*/\sigma^{1/2} = 0.4110 m^{1/2} \text{ m/s}$$

and Equation (7.23) gives the specific endurance as

$$e_t = 9357/m^{3/2} \text{ hours per kg,}$$

from which

$$E_t = -\int_{m_1}^{m_2} e_t \, dm = 9357 \int_{m_2}^{m_1} \frac{dm}{m^{3/2}} = 9.88 \text{ hours.}$$

(b) For $h = 2$ km, $v = 1.0$ and allowing for changes in the mass m the constants are altered to

$$\beta = \beta_m = 0.06$$

and

$$V = 0.411 m^{1/2}/0.76 = 0.5408 m^{1/2} \text{ m/s,}$$

which leads to

$$e_t = 8213/m^{3/2} \text{ hours per kg}$$

and

$$E_t = (8213/9357) \times 9.88 = 8.68 \text{ hours}.$$

(c) A constant speed V with $v = 1.0$ at the start requires first a calculation of V. At the start, $V_e = V_e^* = 109.6 \text{ m/s}$, hence the constant value of V is $109.6/\sigma^{1/2} = 120.9 \text{ m/s}$. At a later time when the mass is m kg, $v = 1.0 \times (50\,000/m)^{1/2}$ and $e_t = 2.206/(\beta m)$ hours per kg. Dividing the flight into four stages as in cases (b) and (c) of Example 7.4, with 2500 kg of fuel used in each stage, the integral of Equation (7.20) can be evaluated and this gives $E_t = 8.14$ hours.

7.10 Effects of climb and descent

So far, the emphasis in this chapter has been on cruise range, although the effects of climb and descent on the total range have been mentioned briefly in § 7.1. The effects are significant because in the climb the specific range r_a is less than it would be in level flight at the same values of W, V and h. In the descent r_a is greater than in level flight but this only partially compensates for the effect of the climb and the overall effect of climb and descent is a loss of range. The *lost range* is defined in such a way that when it is subtracted from the range that is calculated by assuming a level cruise for all the fuel that is available, the result is the true total distance flown, including both the initial climb and the final descent. The *lost time* is defined so that when it is added to the estimated time that would be taken if the whole distance were flown under cruise conditions, the true total flight time is obtained. It is also useful sometimes to consider the concept of *lost fuel*. If this is added to the mass of fuel calculated for a flight by using cruise conditions for the whole distance, the result obtained is the true necessary total mass of fuel. For each of these quantities, lost range, lost time and lost fuel, there is a contribution for the initial climb and another for the final descent. These are calculated separately and the total is obtained by addition.

In estimating the contributions of the climb to the lost range, lost time and lost fuel, comparisons are made between the distance, time and fuel used in the climb and the corresponding quantities which would have been calculated using conditions that exist at the start of the cruise. Similarly, in estimating the contributions of the descent to these lost values, comparisons are made with quantities found for conditions at the end of the cruise. The distinction between the start and end of the cruise is unimportant except for the weight W and the specific range r_a (which is inversely proportional to W), because other quantities such as β, c, f and V do not usually vary much during the cruise but, strictly, all quantities relating to cruise conditions should

Effects of climb and descent

refer to the start of the cruise when climb is considered and to the end when descent is considered.

Holford, Peckham & Cook (1972) have shown that for a typical civil aircraft the lost range is in the region of 100 km, so that errors up to about 10% in calculating its value are of little importance unless the range is very short and this is also true for lost time and lost fuel. Thus it is sufficiently accurate for most purposes to use mean values of quantities such as V, c, f and β in the climb and in the descent and the suffixes CL and D will be used to denote the mean values for these and other quantities such as time and distance in the climb and descent. For the cruise conditions the suffix CR will be used where it is necessary to distinguish quantities in the cruise from those in the climb or descent and as mentioned earlier these should refer to the start of the cruise when climb is considered and to the end when descent is considered. The symbols r_{a1} and r_{a2} will be used to denote the specific range at the beginning and end of the cruise.

In a typical climb to the cruising altitude, both the height and the speed will increase and it is convenient to write the equations in terms of the energy height h_e defined in § 4.5. Assuming that $\cos \gamma = 1$, the rate of increase of energy height in the climb is given by Equation (4.32) as

$$dh_e/dt = V_c' = V_{CL}(f_{CL} - \beta_{CL}), \qquad (7.25)$$

from which the time taken for the climb is approximated as

$$t_{CL} = \Delta h_e / [V_{CL}(f_{CL} - \beta_{CL})], \qquad (7.26)$$

where Δh_e is the total gain of energy height. The distance travelled is then

$$s_{CL} = t_{CL} V_{CL} = \Delta h_e / (f_{CL} - \beta_{CL}). \qquad (7.27)$$

For a turbofan aircraft the mass of fuel used in the climb would be

$$Q_{CL} t_{CL} = F_{CL} c_{CL} t_{CL} \qquad (7.28)$$

and the distance that would be covered at the start of the cruise using the same mass of fuel would be

$$s_{CR} = F_{CL} c_{CL} t_{CL} r_{a1}. \qquad (7.29)$$

The contribution of the climb to the lost range is thus

$$\Delta s_{CL} = s_{CR} - s_{CL} \qquad (7.30)$$

and if it is assumed that the mass of fuel used in the climb is small in comparison with the total aircraft mass, so that the mean weight in the climb is the same as the weight at the start of the cruise, Equations (7.1), (7.26), (7.27), (7.29) and (7.30) show that the lost range can be expressed as

$$\Delta s_{CL} = [\Delta h_e/(f_{CL} - \beta_{CL})]$$
$$\times [(c_{CL}/c_{CR})(V_{CR}/V_{CL})(f_{CL}/\beta_{CR}) - 1]. \quad (7.31)$$

For a propeller aircraft with the specific fuel consumption expressed as $c' = Q/P_e$ the symbols c in Equation (7.31) should be replaced by $c'V/\eta_{PR}$. If η_{PR} is assumed to be the same for the climb as for the cruise the equation becomes

$$\Delta s_{CL} = [\Delta h_e/(f_{CL} - \beta_{CL})][(c'_{CL}/c'_{CR})(f_{CL}/\beta_{CR}) - 1]. \quad (7.32)$$

If the climbing speed V_{CL} is chosen to give maximum V_c' the time t_{CL} is minimum and if F_{CL} and c_{CL} are assumed to be independent of speed the distance s_{CR} is also minimum, but this is not the condition for minimum lost range in the climb. It is found that an increase of V_{CL} above the value giving maximum V_c' gives a useful increase of s_{CL} but a disproportionately small increase of t_{CL} and thus of s_{CR} (Equation (7.29)).

The time that would be taken under cruise conditions to cover the distance s_{CL} that is actually travelled during the climb is

$$t_{CR} = s_{CL}/V_{CR} = \Delta h_e/[V_{CR}(f_{CL} - \beta_{CL})], \quad (7.33)$$

where V_{CR} refers to the start of the cruise. The contribution of the climb to the lost time is therefore

$$\Delta t_{CL} = t_{CL} - t_{CR} \quad (7.34)$$

and Equations (7.26) and (7.33) show that this can be expressed as

$$\Delta t_{CL} = \Delta h_e[(V_{CR}/V_{CL}) - 1]/[V_{CR}(f_{CL} - \beta_{CL})]. \quad (7.35)$$

The lost fuel is the mass of fuel used in flying under the cruise conditions over a distance equal to the lost range. Thus the lost fuel in the climb is $\Delta s_{CL}/r_{a1}$, where r_{a1} is the specific range for the start of the cruise.

The equations used for calculating the lost range, lost time and lost fuel in the descent are essentially the same as for the climb. Some simplifications can be made by assuming that in the descent the thrust is zero and V_e is the same as for the end of the cruise, so that β_D is equal to β_{CR}, the value of β at the end of the cruise. Then the distance travelled during the descent is

$$s_D = \Delta h_e/\beta_D = \Delta h_e/\beta_{CR}, \quad (7.36)$$

where Δh_e is the loss of energy height, and the time taken is

$$t_D = s_D/V_D = \Delta h_e/(\beta_{CR}V_D). \quad (7.37)$$

With zero thrust the specific fuel consumption cannot be used in the equations and an approximation to the mass of fuel used in the descent is better expressed in terms of a ratio of fuel flow rates as

$$Q_D t_D = (Q_D/Q_{CR})(c_{CR} F_{CR} t_D). \tag{7.38}$$

The cruise distance that would be covered with this mass of fuel is

$$s'_{CR} = Q_D t_D r_{a2}, \tag{7.39}$$

where r_{a2} is the specific range at the end of the cruise. Since $F_{CR} = \beta_{CR} W$, Equations (7.1), (7.37), (7.38) and (7.39) show that

$$s'_{CR} = (Q_D/Q_{CR})(V_{CR}/V_D)(\Delta h_e/\beta_{CR}). \tag{7.40}$$

The lost range in the descent is then

$$\Delta s_D = s'_{CR} - s_D = [\Delta h_e/\beta_{CR}][(Q_D/Q_{CR})(V_{CR}/V_D) - 1] \tag{7.41}$$

and since Q_D is much less than Q_{CR}, while the speeds are similar, the lost range given by this equation is normally negative. This is not surprising, because a comparison is being made between normal cruise conditions, for which the fuel consumption is relatively high, and conditions which could be described as an assisted glide, for which the cost in fuel is very small. Thus range in a glide is easily gained, not lost, in comparison with normal cruise.

Holford et al. (1972) suggest that typical values of (Q_D/Q_{CR}) and (V_{CR}/V_D) are respectively 0.15 and $\frac{4}{3}$ and with these values Equation (7.41) becomes

$$\Delta s_D = -0.8 \Delta h_e/\beta_{CR}. \tag{7.42}$$

The total lost range is $(\Delta s_{CL} + \Delta s_D)$ and subtraction of this from the estimated cruise range gives the total distance flown, including both the climb and the descent.

The lost time in the descent is

$$\Delta t_D = t_D - t'_{CR}, \tag{7.43}$$

where t'_{CR} is the time that would be taken in cruise conditions to cover the distance s_D that is actually travelled during the descent. Thus

$$\Delta t_D = s_D/V_D - s_D/V_{CR} = [\Delta h_e/(\beta_{CR} V_{CR})][(V_{CR}/V_D) - 1] \tag{7.44}$$

and for the same suggested value $V_{CR}/V_D = \frac{4}{3}$ this gives

$$\Delta t_D = \Delta h_e/(3\beta_{CR} V_{CR}). \tag{7.45}$$

The total lost time is $(\Delta t_{CL} + \Delta t_D)$ and addition of this to the time calculated for the whole flight distance under cruise conditions gives the true total flight time. Holford et al. (1972) point out that the simple expression (lost range)/(cruise speed) gives a good approximation to the total lost time, although it is not analytically correct.

The lost fuel in the descent is equal to $\Delta s_\mathrm{D}/r_{a2}$. Addition of this to the lost fuel in the climb gives the total lost fuel, and the total mass of fuel actually used is obtained by adding this total lost fuel to the mass of fuel calculated for the whole flight distance under cruise conditions.

Climb and descent also have effects on endurance and in considering these it will be assumed that the aircraft is required to fly at a specified height and the endurance at this height is to be estimated. For calculation of the effects of climb and descent it is necessary to find only the mass of fuel used in climb and descent, so that this mass can be subtracted from the total mass of fuel that is available. The mass of fuel used in the climb is given by Equations (7.26) and (7.28) and the mass used in the descent is given by Equations (7.37) and (7.38). For maximum endurance at the specified height the speed in the climb should be adjusted so that Q/V_c' always has minimum value, where $V_\mathrm{c}' = \mathrm{d}h_\mathrm{c}/\mathrm{d}t$.

7.11 Effects of engine failure

The effects of engine failure during the take-off have been considered in Chapter 6 and the only case that will be considered here is the failure of one engine of a multi-engine aircraft during the cruise. It will be assumed at first that all the engines are operating at their cruise rating before the failure occurs, so that no additional thrust can be obtained from the operating engines to compensate for the loss of thrust of the failed engine.

Except when the failed engine is the central one of a three-engine aircraft, failure of an engine means that the aircraft has to fly with more thrust on one side than the other, since it is usually unacceptable to decrease the total thrust still further (on a four-engine aircraft) by shutting down another engine on the opposite side to that of the failed engine. The asymmetric thrust produces a yawing moment, which must be counteracted by deflection of the rudder, and this leads to some increase of drag. Deflection of the ailerons is also required and in general there is a small angle of sideslip and also a small angle of bank. Torenbeek (1982) in Chapter 9 and Appendix G explains the various factors contributing to the additional drag in asymmetric flight and notes that for turbofan aircraft the most important is the lift-dependent drag of the fin, although for propeller aircraft there can be large effects associated with the asymmetric slipstream. Torenbeek (1982) gives equations for calculating the additional drag for a turbofan aircraft, representing a small addition to the loss of thrust caused by the failure of one engine, and discusses the optimum values of sideslip and bank angle. A further factor to be considered is the

Effects of engine failure

drag of the failed engine, but for a turbofan this is usually small and even with a propeller the drag can be kept to a low value by 'feathering' the propeller.

With all the engines operating initially at their cruise rating, a failure of one engine makes it necessary to reduce the cruising height in order to increase the thrust or power of the remaining engines and there must then be either a reduction of V, or an increase of v above the normal cruising value, or a combination of both these changes. For a turbofan aircraft the specific range r_a is given by Equation (7.1) and if c is constant it is directly proportional to V/β, for a given weight. An increase of v above the optimum value will increase β, so that either this change or a reduction of V, or both, will lead to a reduction of r_a.

EXAMPLE 7.7. REDUCTION OF RANGE OF TURBOFAN AIRCRAFT CAUSED BY ENGINE FAILURE

The aircraft described in Example 7.3 has four engines and is cruising at a constant true air speed of 236 m/s, with v held constant at 1.15 (condition I(b) of Example 7.3). The engines are all operating at the rated cruise thrust and when the mass has decreased to 85 000 kg one of the outer engines fails. The additional drag caused by asymmetric flight increases the value of K_1 in Equation (3.6) by $0.1 \Delta F/(\tfrac{1}{2}\rho_0 V_e^2 S)$, where ΔF is the excess thrust on one side of the aircraft. After the engine failure a cruise-climb is resumed, at a reduced height, and it may be assumed that for each engine the cruise rated thrust is proportional to $\sigma^{0.6}$. Find the reduction of range caused by the engine failure if there is no change in the EAS when the engine fails.

Example 7.3 shows that the conditions immediately before the engine failure are

$$V_e^* = 122.54 \times (0.85)^{1/2} = 112.98 \text{ m/s}, \qquad \beta = 0.0624,$$
$$V_e = 1.15 V_e^* = 129.93 \text{ m/s}$$

and

$$\sigma = 0.3566 \times 0.85 = 0.3031,$$

for which $h = 10.84$ km.

A first approximation to the required value of σ, immediately after the engine failure, can be obtained by neglecting the additional drag caused by asymmetric flight. Then V_e^*, V_e, v and β are all unchanged and there is no change in the required thrust, which is equal to $D = \beta W$, but this must be provided by three engines instead of four. The thrust of each engine must be increased in the ratio 4/3 and this requires a new value of σ given by

$$(\sigma/0.3031)^{0.6} = \tfrac{4}{3}$$

and hence $\sigma = 0.4896$, for which $h = 6.86$ km. At this height the rated cruise thrust of *one* engine must be

$$\beta W/3 = 17.34 \text{ kN}$$

and this is the excess thrust ΔF on one side of the aircraft, from which the increase of K_1 is found as

$$\Delta K_1 = 0.1 \times 17.34 \times 10^3 / [\tfrac{1}{2} \times 1.225 \times (129.93)^2 \times 160]$$
$$= 0.00105.$$

The new value of K_1 is therefore 0.02105 and $\beta_m = 0.0615$. Equation (3.17) shows that $V_e^* \propto K_1^{-1/4}$, for constant W and K_2, and thus

$$V_e^* = 112.98(0.02/0.02105)^{1/4} = 111.54 \text{ m/s}.$$

With no change of V_e, the speed ratio v is

$$129.93/111.54 = 1.165$$

and hence

$$\beta = \tfrac{1}{2} \times 0.0615(v^2 + v^{-2}) = 0.0644.$$

The ratio in which the thrust of each engine must be increased after the engine failure is now not simply 4/3, but

$$4/3 \times (0.0644/0.0624) = 1.376$$

and a corrected value of σ is obtained by equating this ratio to $(\sigma/0.3031)^{0.6}$, leading to

$$\sigma = 0.5160, \quad \text{for which} \quad h = 6.39 \text{ km}.$$

The revised value of ΔF is $17.34 \times (0.5160/0.4896)^{0.6} = 17.90$ kN. Hence $\Delta K_1 = 0.00108$ and the change from the earlier value of K_1 is insignificant. Using the revised value of σ, the conditions in the cruise at reduced height, after the engine failure, are found as $V = V_e \sigma^{-1/2} = 129.93 \times (0.516)^{-1/2} = 180.9$ m/s and since $a/a_0 = \theta^{1/2} = 0.9251$,

$$M = 0.5746, \quad c = 1.586 \times 10^{-5} \text{ kg/N s}$$

and

$$V/(gc\beta) = 18\,053 \text{ km}.$$

Then Equation (7.3) shows that the range for this part of the cruise, after the engine failure is

$$18\,053 \ln(0.85/0.7) = 3505 \text{ km}.$$

As shown in Example 7.3, the range for the first part of the cruise, before the engine failure, would be

$$22\,208 \ln(1/0.85) = 3609 \text{ km}.$$

Thus the total range is 7114 km, a reduction of 831 km due to the engine failure.

It is also worth noting that without an engine failure the total flight time is

$$7945/(236 \times 3.6) = 9.35 \text{ hours},$$

whereas when engine failure occurs the total time is

$$(3609/236 + 3505/180.9)/3.6 = 9.63 \text{ hours}.$$

Thus the flight time for the reduced range, after an engine failure, is slightly greater than the time for the full range without a failure.

The procedure described here, keeping V_e unchanged, is of course not the only possibility. There must always be a reduction of height, but the loss of range can often be reduced by allowing some increase of V_e, giving greater values of v, β and c, but probably a reduction of $c\beta/V$.

For a turboprop aircraft the specific range is given by Equation (7.6) and is proportional to $\eta_{PR}/(c'\beta)$, for a given weight. Reduction of speed at constant v has no effect on r_a unless it causes a change of η_{PR}/c'. Equation (5.24) shows that η_{PR}/c' is approximately proportional to $M^{0.1}$, so that a small reduction of r_a may be expected as V decreases. If there is an increase of v as the height is reduced after an engine failure this will lead to an increase of β and a loss of r_a, but the loss of r_a after an engine failure is usually less than it is for a turbofan aircraft.

In some cases, particularly when the installed thrust is determined mainly by the take-off and climb requirements, the engines may be operating with the thrust well below the cruise rated value when one engine fails. It will then be possible to increase the thrust of the engines that are still operating and it will sometimes be possible to maintain the original cruising height and speed. Even if some loss of height has to be accepted, the loss will be less than it was in the case considered earlier and hence the loss of range will be less. In some extreme cases the thrust in normal cruising flight is so far below the rated value that the increase of thrust needed when one engine fails leads to a substantial reduction of sfc, because of the effects described in § 5.2.3 and 5.4.2. This may lead to an increase of specific range when one engine fails as shown, for example, by some data given by Dornier GmbH (1986) for the Dornier 228 aircraft with twin turboprop engines. For this aircraft the maximum specific range is increased when one engine is shut down, by about 35% at a height of 1.5 km and by about 22% at 4.5 km.

7.12 Effects of wind

In all the preceding discussion of range it has been assumed that the aircraft is flying in still air. It is obvious that when there is a

208   Fuel consumption, range and endurance

headwind or tailwind the range and specific range measured relative to the ground will be different from the values in still air, but it is less obvious that a wind also has an effect on the air speed required to give maximum specific range relative to the ground.

It will be assumed at first that a wind of velocity V_w blows along the track of the aircraft, with zero components of velocity both in the vertical direction and horizontally normal to the track. (Reference may be made to ESDU 73018 (1980) for some effects of these other components.) For consistency with Chapter 6 the velocity V_w is taken to be positive when it is a headwind and the specific *ground* range is then

$$r_g = -dR_g/dm = (V - V_w)/Q, \quad (7.46)$$

where Q is the fuel flow rate and R_g is the range of the aircraft relative to the ground. If r_g can be estimated for each instant of the flight the ground range R_g can be found from

$$R_g = -\int_{m_1}^{m_2} r_g \, dm, \quad (7.47)$$

where m_1 and m_2 are the masses at the beginning and end of the flight.

To give an indication of the effects of wind on the conditions for maximum r_g the value of v for these conditions will be found for one particular case, a cruise at a specified fixed height, making several different assumptions about specific fuel consumption. It is convenient to introduce a dimensionless *wind speed ratio* defined as

$$v_w = V_w/V^*,$$

where $V^* = V_e^*/\sigma^{1/2}$, the true air speed for minimum β in straight and level flight. For a turbofan aircraft in level flight $Q = cF = c\beta W$ and Equation (7.46) becomes

$$r_g = (V - V_w)/(c\beta W). \quad (7.48)$$

If c is assumed to be independent of V, for the given height, r_g is maximum for a given weight when $\beta/(V - V_w)$ is minimum. Equation (3.18) shows that

$$\beta/(V - V_w) = \tfrac{1}{2}\beta_m[v^2 + v^{-2}]/[V^*(v - v_w)]$$

and the quantity that is required to be minimum is

$$(v^2 + v^{-2})/(v - v_w),$$

the condition for this being

$$v_w = \tfrac{1}{2}v[1 - 2/(v^4 - 1)] \quad (7.49)$$

and this is plotted as curve I in Figure 7.7. For $v_w = 0$, $v = 1.316$ as given by Equation (7.17).

Effects of wind

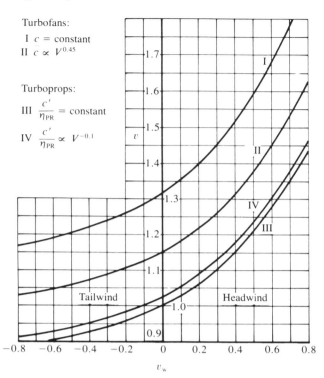

Figure 7.7. Effect of wind speed on speed ratio v for maximum specific range at constant height.

If the specific fuel consumption c varies with speed as given by Equation (5.11), $c \propto v^n$ and the condition for maximum r_g is that $v^n \beta/(V - V_w)$ should be minimum. Thus the quantity that is required to be minimum is

$$[v^{(n+2)} + v^{(n-2)}]/[v - v_w]$$

and the condition for this is

$$v_w = v\{1 - [1 + v^4]/[(n+2)v^4 + (n-2)]\}, \qquad (7.50)$$

which is plotted as curve II in Figure 7.7 for $n = 0.45$. The value of v for $v_w = 0$ is now 1.152 in accordance with Equation (7.18).

For a turboprop aircraft $Q = c'FV/\eta_{PR} = c'\beta WV/\eta_{PR}$ and

$$r_g = \eta_{PR}(V - V_w)/(c'\beta WV). \qquad (7.51)$$

If c'/η_{PR} is assumed to be independent of speed the condition for maximum r_g is that $\beta V/(V - V_w)$ should be minimum. Thus the quantity to be minimised is

$$v(v^2 + v^{-2})/(v - v_w)$$

and the condition for this is

$$v_w = 2v(v^4 - 1)/(3v^4 - 1), \tag{7.52}$$

which is plotted as curve III in Figure 7.7. It happens that this curve also gives the condition for minimum angle of glide relative to the ground, with zero thrust. The reason for this is that with zero thrust the aircraft travels a distance V relative to the air in unit time and loses βV in height. However, in the presence of the wind the distance travelled relative to the ground is $(V - V_w)$, so that the angle of glide relative to the ground is $\beta V/(V - V_w)$ and the condition for this to be minimum is given by Equation (7.52).

As explained in § 5.4.2 there is usually a slight decrease of c'/η_{PR} for a turboprop as the speed increases and this can be represented by Equation (5.24). If it is assumed that $c'/\eta_{PR} \propto V^{-n}$ the condition for maximum r_g is that $\beta V^{(1-n)}/(V - V_w)$ should be minimum. Thus the requirement is that $v^{(1-n)}(v^2 + v^{-2})/(v - v_w)$ should be minimum and the condition for this is

$$v_w = v\{1 - [1 + v^4]/[(3-n)v^4 - (1+n)]\}. \tag{7.53}$$

With $n = 0.1$ this is plotted as curve IV in Figure 7.7, showing that there is only a small difference between $n = 0$ (curve III) and $n = 0.1$ (curve IV).

When the wind direction is not parallel to the track of the aircraft the value of v for maximum r_g can be found from the equations and graphs given in ESDU 73018 (1980), for either constant c or constant c'/η_{PR}. The unbroken curves in Figure 7.8 show these optimum values of v for constant c, for wind directions at 45° to the headwind or tailwind direction. The abscissa in this figure is the ratio of the wind speed to V^* and is denoted by v'_w instead of v_w, to indicate that it refers to a wind in any specified direction.

The broken curves in Figure 7.8 show the results obtained from Equation (7.49), making the simple assumption that the optimum v depends only on the component of wind velocity along the track of the aircraft. The differences between the broken and unbroken curves are quite small for values of v'_w up to about 0.25, showing that unless the wind is very strong and V^* is small it is sufficiently accurate to estimate the optimum value of v from the component of wind velocity along the aircraft track.

Variation of payload with range

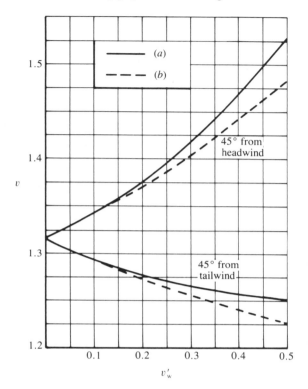

Figure 7.8. Speed ratio v for maximum specific range at constant height (c = constant). (a) Exact result from ESDU 73018 (1980). (b) Calculated from component along aircraft track, using Equation (7.49).

7.13 Variation of payload with range

The total mass of an aircraft at take-off is limited to a specified maximum value by the structural design and by the requirements of take-off, in particular the need for clearance of the screen height and attainment of an adequate angle of climb after failure of one engine (see § 6.5). The structural limitation is dependent on assumptions made at an early stage in the design, essentially related to the average weight of each passenger and the average density of baggage and freight in the holds. These assumptions determine the mass distribution throughout the fuselage and have a direct influence on the structural design; for example they are important in determining the bending moments at the wing roots during flight. An important aim in making all these design assumptions is to define eventually the maximum allowable payload and also to show how the combined mass of payload and fuel must be limited.

212   *Fuel consumption, range and endurance*

It is usual to represent the limitations on payload, with implied limitations on fuel load, by constructing a payload–range diagram of the kind shown in Figure 7.9. It is important to recognise that in discussing this diagram 'fuel' means the fuel that is actually used, sometimes called the 'stage fuel', because the weight of the reserve fuel and any other fuel that is not normally used must be taken to be part of the basic aircraft weight. The diagram should be accompanied by a specified flight profile or a specified cruise procedure, such as a stepped cruise at stated speeds and heights, and the value of the diagram is primarily for comparing the capacities of similar aircraft, subject to comparable flight specifications, or for demonstrating the commercial potential of a particular aircraft under precisely specified conditions.

The point A in Figure 7.9 shows the maximum payload with zero fuel and represents the maximum allowable addition to the operating weight empty (OWE), while keeping within the structural limitations already mentioned. Proceeding along the line AB, only the fuel load is increased as the range is extended from zero, the fuel going at first into wing tanks because of the beneficial effect of this on the wing root bending moment. At the point B the maximum take-off weight (MTOW) is reached, this being the sum of the OWE, the maximum payload and as much fuel as can be added without exceeding the MTOW.

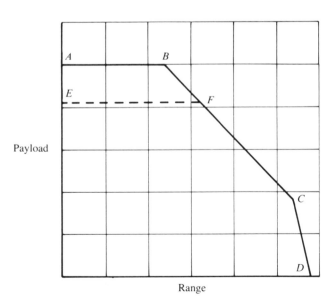

Figure 7.9. Typical payload–range diagram.

Variation of payload with range

The point *B* represents the maximum range for the maximum payload and in moving from *B* along the line to *C* there is a necessary reduction of payload as more fuel is added to gain additional range, since the total weight of (fuel + payload) cannot be allowed to exceed the value at *B*.

At the point *C* all the fuel tanks are full, often including some in the fuselage as well as those in the wing, and the total aircraft weight is still at the MTOW. A further reduction of payload allows a modest increase in range along the line *CD*, but only because the reduction of total weight increases the specific range as shown by Equation (7.1). For a civil aircraft a line such as *EF* is often included, to represent the payload when there is a full load of passengers and their associated baggage. The allowable additional payload shown by the difference between *AB* and *EF* can be made up with freight.

8

Turning performance

In earlier chapters, except in the discussions of the landing flare and the take-off transition immediately after lift-off, it has been assumed that the flight path is straight, so that there is no component of acceleration normal to the flight path. In this chapter flight in a curved path will be considered, concentrating on the usual form of banked turn as shown in Figure 8.1, in which the angle of bank is adjusted so that there is no sideslip and therefore no component of aerodynamic force normal to the plane of symmetry of the aircraft. In such a turn the required lift is greater than the weight, thus C_L is greater than it would be in straight and level flight at the same speed and consequently the drag is also greater. This raises the requirement for thrust, even to maintain level flight, and thus the rate of climb obtainable with the maximum available thrust is reduced and may become negative. As the turn becomes tighter and the normal acceleration V^2/R is increased, due to either a high speed or a small turn radius, or both, there will be increased demands for C_L and for thrust to maintain height, with the consequence that limitations may be imposed by stalling or buffeting or by the engine rating.

This chapter addresses the interdependences among speed, rate of turn, rate of climb and additional 'g-load' on the pilot, as well as the limitations on one or other of these when some are fixed. The stalling and buffet boundaries are shown to become more restrictive as the height increases and the concept of flight ceiling, introduced in § 2.6, is seen to be strongly influenced by turning, because the increase of drag in a turn reduces the maximum height which can be maintained by the available thrust.

8.1   Curved flight in a vertical plane

Before starting the discussion of banked turns it is useful to introduce the subject of flight in a curved path by considering the simple case in which the flight path is curved only in the vertical plane

Curved flight in a vertical plane

Figure 8.1. The British Aerospace Experimental Aircraft Programme (EAP) demonstrator during handling trials in 1987.

and the plane of symmetry of the aircraft remains vertical. This is the form of flight path used in a pull-out from a dive and in a loop. At any instant the flight path has a radius of curvature R, taken to be positive when the normal acceleration V^2/R is in the same direction as the lift force. The forces on the aircraft are as shown in Figure 2.1 but an additional term must be inserted in Equation (2.1) to represent the force needed to generate the normal acceleration V^2/R. The revised equation is

$$mV^2/R = L - mg \cos \gamma \tag{8.1}$$

and this is independent of the force and acceleration components along the flight path.

A ratio of great importance in the study of curved flight is the *load factor*

$$n = L/W \tag{8.2}$$

and with $L = nW$ Equation (8.1) shows that the acceleration normal to the flight path is

$$V^2/R = L/m - g \cos \gamma = g(n - \cos \gamma). \tag{8.3}$$

Since $\cos(-\gamma) = \cos \gamma$ the Equations (8.1) and (8.3) are unaffected by a change in the sign of γ, i.e. they are equally valid for upward and downward portions of the flight path. It should also be noted that the equations are valid for any form of curved path in a vertical plane, since the radius of curvature R may vary in any way along the flight path. An example is the loop manoeuvre shown in Figure 8.2, in which the aircraft usually gains speed in a dive before entering the loop, then departs from the straight dive at a point A and reaches the condition of level flight at B, still with a high speed. A value of n greater than 1 is usually maintained over most of the flight path from A to the end point D, although a value between 0 and 1 is acceptable in the region near C. At the point C the flight path is horizontal with the aircraft inverted but if $n > 0$ the pilot still feels that he is being pressed into his seat and not tending to fall out. Since $\cos \gamma = -1$ at C, Equation (8.3) shows that if $n > 0$ the acceleration V^2/R is greater than g and this is consistent with a downward force exerted by the seat on the pilot. As the aircraft climbs from B to C there is usually a substantial reduction of speed and a decrease of radius of curvature, so that the flight path is not circular.

A pull-out from a dive with $n = 2$ (for example) is usually described as a '2g pull-out' because with this value of n the force component normal to the flight path, acting on the aircraft or on the pilot, is equal to twice the weight of either. Thus the force between the pilot and his seat has double the normal value and he feels as if g had been

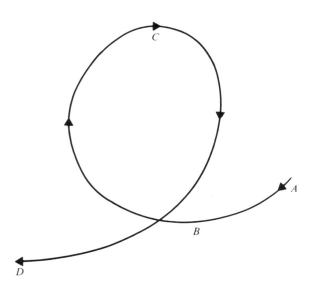

Figure 8.2. Loop manoeuvre.

multiplied by 2. Equation (8.3) shows that n is *not* a true measure of acceleration except in the particular case where $\gamma = \pm 90°$ and for this reason n should not be called an acceleration factor, although that term is sometimes used.

The meaning of n can be explained further by considering the instrument that is usually called an accelerometer, although when it is mounted in the usual way in an aircraft, with its sensitive axis normal to the flight path, it measures n and not acceleration. The instrument consists in principle of a mass that is restrained by a spring and mounted so that it can move only along a fixed axis. When the axis is vertical and the instrument is at rest the force on the mass (apart from the spring force) is equal to its weight and the scale is graduated so that in this condition it reads 1. When the force is n times the weight of the mass the scale reads n, so that when the instrument is mounted in an aircraft with its axis normal to the flight path it reads the load factor directly.

The maximum value of n that can safely be applied by the pilot may be limited either by human factors, by the structural strength of the aircraft or by the possibility of stalling or buffeting. These limitations will be discussed in § 8.3 and 8.4.

8.2 Equations for a banked turn

A *true-banked* turn is one in which there is no sideslip, i.e. no component of relative air velocity normal to the plane of symmetry of the aircraft and as a consequence there can be no component of aerodynamic force normal to the plane of symmetry. As mentioned earlier this is the usual form of banked turn and throughout this book the simpler expression 'banked turn' will be used with this meaning.

For simplicity it will be assumed at first that the aircraft is flying in a horizontal path. Figure 8.3 shows the aircraft as seen from the front, with the lift force L acting in the plane of symmetry and normal to the flight path. The angle ϕ between the plane of symmetry and the vertical is known as the *angle of bank*. Since the acceleration normal to the flight path is horizontal and equal to V^2/R, where R is the radius of curvature of the flight path, the equations defining the motion are

$$mg = L \cos \phi \tag{8.4}$$

and

$$mV^2/R = L \sin \phi. \tag{8.5}$$

Hence

$$\tan \phi = V^2/(gR) \tag{8.6}$$

218 *Turning performance*

and

$$n = L/W = \sec \phi = (1 + \tan^2 \phi)^{1/2} = [1 + V^4/(gR)^2]^{1/2}. \qquad (8.7)$$

The acceleration normal to the flight path and in the horizontal plane is

$$V^2/R = g \tan \phi = ng \sin \phi \qquad (8.8)$$

and again it can be seen that in general n is not a true measure of acceleration, athough a nomenclature similar to that used for a pull-out from a dive is commonly used and a turn with $n = 2$ (for example) is described as a '2g turn'. For $n = 2$ the Equations (8.7) and (8.8) show that $\phi = 60°$ and the inward acceleration is $3^{1/2}g$, but as n becomes large $\phi \to 90°$ and $V^2/R \to ng$.

Equation (8.7) shows that for a given value of n the radius R is proportional to V^2 and so becomes very large at high speeds unless n is large, but the maximum value of n that can be applied by a pilot is limited in practice by several factors which are considered in § 8.3, 8.4 and 8.6.

8.3 Structural and human limitations on the load factor

For a combat aircraft the ability to turn with a small radius R is a valuable quality and this requires a large value of the load factor. For this reason these aircraft are usually designed with sufficient strength to allow a load factor as high as about 8. For transport aircraft the load factors required by certification for turns and for other manoeuvres are quite small and the limiting structural strength is dictated mainly by the loads imposed by gusts, but it is usually found that when the gust load requirement is satisfied there is sufficient strength to allow for a load factor up to about 2.5 or 3.

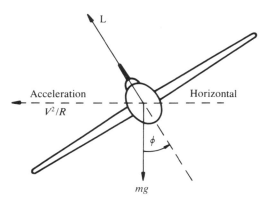

Figure 8.3. Aircraft in a banked turn, as seen from the front.

There are important adverse effects of high load factors on a man in an aircraft, since his apparent weight is increased to n times the normal weight. This causes a draining of blood from the head to the lower part of the body and may lead to 'black-out' (loss of vision) or even loss of consciousness. Military pilots flying combat aircraft wear 'anti-g' suits (sometimes called g suits) to increase their tolerance to high load factors and with these suits values of n as high as the structural limit of about 8 can be sustained without blacking out. The effects of high n on a pilot without an anti-g suit have been summarised by Tobin (1945). Black-out normally occurs when n is between 4.5 and 5, but higher values can be tolerated without ill effects for very short periods. Passengers in commercial aircraft are much less tolerant than military pilots and except in emergencies they expect to feel no lateral acceleration, thus requiring all turns to be true-banked, and in any turn they expect the normal load factor n to be kept low enough to be hardly noticeable. Civil transport aircraft in normal service do not usually exceed a bank angle of about 30°, for which $n = 1.15$.

8.4 Turning limitations due to stalling or buffeting

For an aircraft of given weight, flying at a fixed height and a constant speed V, the lift coefficient is directly proportional to the load factor n and an upper limit of n is reached when stalling occurs and $C_L = C_{Lmax}$. If C_L is the lift coefficient in the turn and C_{L0} is the lift coefficient at the same speed and height in straight and level flight (with $n = 1$), then

$$C_L = nC_{L0}. \tag{8.9}$$

Defining a speed ratio N as V/V_s, where V_s is the stalling speed in straight and level flight at the same height,

$$C_{L0} = C_{Lmax}/N^2 \tag{8.10}$$

and in the turn

$$C_L = (n/N^2)C_{Lmax}. \tag{8.11}$$

In the limiting case where $C_L = C_{Lmax}$ in the turn, $n = N^2$ and this gives the maximum possible value of n for any given speed.

In this simple analysis it has been assumed that C_{Lmax} is independent of speed. The assumption is correct at low Mach numbers but as M increases above about 0.3 or 0.4 there is often a reduction of C_{Lmax} and moreover at Mach numbers above about 0.6 or 0.7 the highest usable value of C_L may be limited not by conventional stalling but by buffeting, caused by unsteady separated flow induced by a shock wave. Buffeting has been studied extensively in wind tunnels and in flight and

Mabey (1973) has defined criteria in terms of a buffeting coefficient for 'light', 'moderate' and 'heavy' buffeting. Using these criteria Langfelder (1974) has given values of C_L over a range of Mach number for the three levels of buffeting for a wing with 45° sweepback. Figure 8.4 is based on Langfelder's data and shows both the level of C_L for moderate buffeting and the maximum value of C_L for the same wing. As in many other cases there is a large difference for $M > 0.6$ between C_{Lmax} and C_L at the buffet limit, but in this range of M it is only the buffet limit that is relevant in considering turning performance.

The turning performance of an aircraft as given by Equation (8.7) will now be examined, taking into account the limitations imposed by stalling and buffeting. Especially at high altitudes the limitations become important when a high rate of turn is required, i.e. a large load factor. The discussion is therefore relevant mainly to combat aircraft for which both the aircraft structure and the crew can survive load factors up to about 8 if anti-g suits are worn. For transport aircraft, turns with load factors above about 2 are most unlikely.

As an example, an aircraft will be considered which has a wing loading of 4 kN/m^2 and stalling and buffeting characteristics broadly similar to those shown in Figure 8.4. For simplicity it will be assumed that the boundary for moderate buffeting is represented by the constant value $C_L = 0.75$ for $0.7 < M < 0.9$. The maximum lift

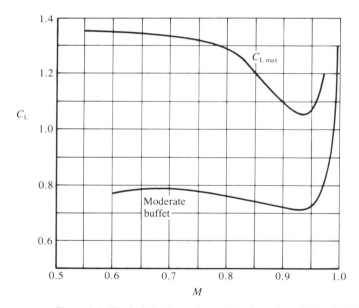

Figure 8.4. Typical limiting values of C_L, from Langfelder (1974).

coefficient will be taken to have the constant value of 1.35 for all Mach numbers up to 0.7 but in order to provide a margin of safety from stalling it will be assumed that C_L must not be allowed to exceed 1.0. (No corresponding margin from the moderate buffet limit is needed because aircraft can usually be flown safely, if uncomfortably, even with severe buffeting.)

Equation (8.7) shows that the relation between n, R and V is independent of height and wing loading and this relation is shown by the full lines in Figures 8.5 and 8.6, which refer to heights of 0 and 10 km respectively, both figures giving the same R for any specified values of V and n. (The Mach number for given values of n, R and V increases with height because the velocity of sound decreases.) The broken lines in the two figures show the limits imposed by the assumed maximum allowable lift coefficients, 1.0 for $M < 0.7$ and 0.75 for $M > 0.7$, for the specified values of height and wing loading. As $n \rightarrow 1$, $R \rightarrow \infty$ and this is shown by the curves for $C_L = 1$ as V decreases towards the value giving $C_L = 1$ in straight and level flight, corresponding to Mach numbers of 0.237 at $h = 0$ and 0.465 at $h = 10$ km.

The assumption made here about the limiting values of C_L implies that if the Mach number is increased gradually at a constant C_L of (say) 0.85 buffeting will start suddenly as M passes through 0.7. In reality the buffeting is likely to build up gradually over a band of Mach number and the behaviour implied by Figures 8.5 and 8.6 must be regarded as a simple approximation to the true situation. To illustrate the effects of the assumed limits of C_L on the turning performance three cases will be considered for a height of 10 km. In the first case the Mach number is increased gradually from 0.66, while maintaining a constant load factor of 2.0, and Figure 8.6 shows that buffeting will start when $M = 0.7$ and $R = 2.6$ km. The only way to increase the speed further without buffeting is to reduce the load factor to 1.7 or less. In the second case the Mach number is reduced gradually from 0.85, while keeping n at the constant value of 2.0; buffeting will start at $M = 0.76$ and this can be suppressed either by increasing M again or by reducing n. In the third case the speed is allowed to fall gradually, while maintaining a constant turn radius of 3.0 km, and buffeting will start when M has fallen to 0.775 and $n = 2.1$. The buffeting can be suppressed either by increasing the radius of the turn, thus reducing n, or by restoring the speed to a higher value.

The broken lines for constant C_L in Figures 8.5 and 8.6 show how the minimum radius of turn varies with M, V or n. With the assumption made here the minimum radius increases discontinuously as M rises through 0.7 and buffeting starts if $C_L > 0.75$. Both of the broken lines in Figure 8.5 and the one for $C_L = 0.75$ in Figure 8.6 show that for high speeds, for which large values of n are possible, the

222 *Turning performance*

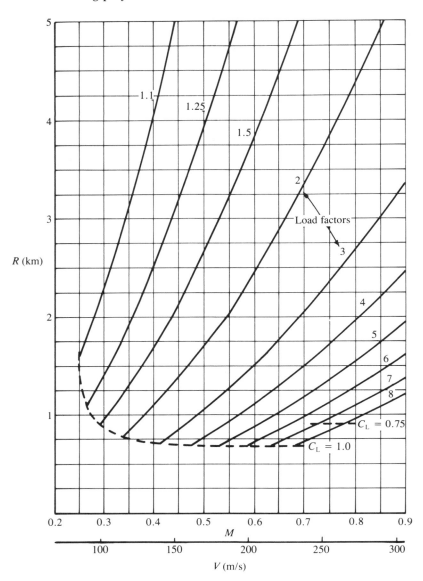

Figure 8.5. Radius of turn at sea level, limited by stall or buffet ($w = 4 \text{ kN/m}^2$).

Turning limitations due to stalling or buffeting

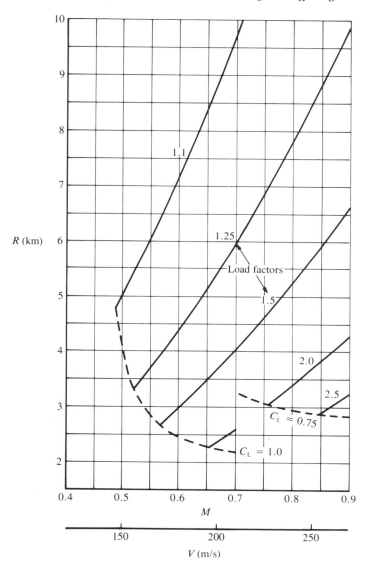

Figure 8.6. Radius of turn at $h = 10$ km, limited by stall or buffet ($w = 4$ kN/m^2).

minimum radius is nearly independent of M (or V) and n. This can be explained by using a rearrangement of Equation (8.7), namely

$$R = V^2/[g(n^2 - 1)^{1/2}], \tag{8.12}$$

for if it is noted that

$$L = nW = \tfrac{1}{2}\rho_0 \sigma C_L S V^2,$$

from which the speed term can be expressed as

$$V^2 = 2nw/(\rho_0 \sigma C_L), \tag{8.13}$$

then the turning radius is given by the more revealing equation

$$R = 2nw/[\rho_0 \sigma C_L g(n^2 - 1)^{1/2}]. \tag{8.14}$$

This shows that for constant C_L, w and σ

$$R \propto n(n^2 - 1)^{-1/2}$$

and for values of n greater than about 3 this function of n decreases only slightly as n increases. Equation (8.14) also shows that for constant C_L and large values of n the minimum radius becomes approximately proportional to w/σ. Thus for a given limiting C_L the minimum radius of turn becomes very large at great heights, unless the wing loading is low. This has important implications for the manoeuvrability of combat aircraft at high altitudes.

The *rate of turn* is the angular velocity about a vertical axis

$$\Omega = V/R \tag{8.15}$$

and Equations (8.7) and (8.8) show that

$$\Omega = g(n^2 - 1)^{1/2}/V. \tag{8.16}$$

Figures 8.7 and 8.8 show the rate of turn in radians per second plotted against M (and V) for a range of load factors. As for Figures 8.5 and 8.6 the relation between n, Ω and V is independent of height and wing loading and the information given by the full lines in Figures 8.7 and 8.8 is essentially the same as that shown in Figures 8.5 and 8.6, but the lines of constant C_L now represent upper limits of Ω and n for any given speed and for the specified values of height and wing loading. (The broken lines marked with values of ϵ refer to thrust limitations to be discussed in §8.6.) The rate of turn is of course zero for $n = 1$, giving $R = \infty$ as mentioned earlier.

Figure 8.9 shows the maximum load factor as a function of Mach number, derived either from Figures 8.5 and 8.6 or from Figures 8.7 and 8.8. The discontinuity shown at $M = 0.7$ is of course a consequence of the simple assumption that has been made about maximum allowable values of C_L. In reality there would be some smoothing of

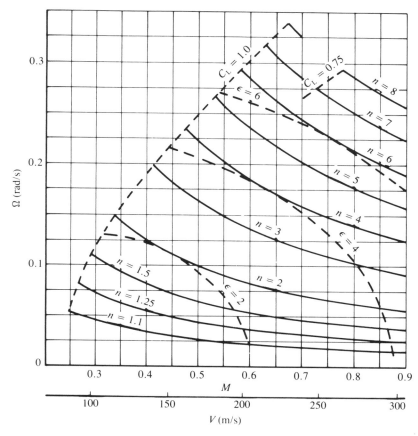

Figure 8.7. Rate of turn at sea level, limited by stall, buffet or thrust ($w = 4 \text{ kN/m}^2$). Lines of constant ϵ show conditions for $P_{se} = 0$, assuming that $C_L^* = (\frac{1}{3})^{1/2}$.

this change, but a reduction of maximum n is to be expected as the Mach number rises through values in the region of 0.7.

The lines of constant C_L representing the limiting conditions in Figures 8.5–8.9 are calculated for the assumed wing loading of 4 kN/m^2 and for the stated values of height and C_L, but Equations (8.13) and (8.14) show that they also apply to any other combination of height, w and C_L giving the same value of the ratio $w/(\sigma C_L)$. For example, the lines shown in the figures for $C_L = 1$ apply also to an aircraft at the same height with $w = 3 \text{ kN/m}^2$ and $C_L = 0.75$. In contrast, the curves in Figures 8.5–8.8 relating the *speed* to n, R and Ω apply to any aircraft with any wing loading at any height, although the relation shown between speed and Mach number will change for a different choice of height.

226   *Turning performance*

In general, the upper limit of C_L may vary continuously with Mach number in any prescribed way. The lines representing maximum values of n and Ω and minimum values of R may then be drawn on diagrams like Figures 8.5–8.8 by drawing first some lines of constant C_L, as already shown, and then choosing the appropriate limiting C_L at each value of M to obtain the limiting curve on each diagram.

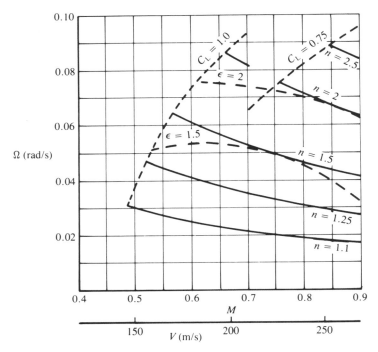

Figure 8.8. Rate of turn at $h = 10$ km, limited by stall, buffet or thrust ($w = 4$ kN/m^2). Lines of constant ϵ show conditions for $P_{se} = 0$, assuming that $C_L{}^* = (\frac{1}{3})^{1/2}$.

EXAMPLE 8.1. MAXIMUM LOAD FACTOR, LIMITED BY STALLING OR BUFFETING

An aircraft with a wing loading of 3.5 kN/m² makes banked turns at a height of 5 km. Find the maximum possible load factor (a) for $V = 100$ m/s if C_L is not to exceed 1.2 and (b) for $V = 200$ m/s if C_L is not to exceed 0.8. In each case also find the angle of bank, the radius of the turn, the rate of turn and the time taken for one complete turn, when n has the maximum value.

The lift coefficient in straight and level flight is

Turning limitations due to stalling or buffeting

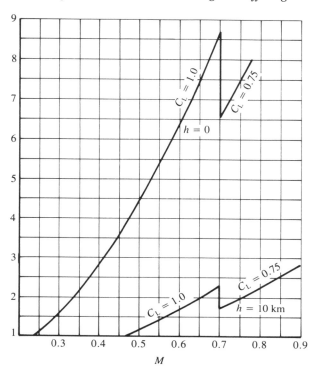

Figure 8.9. Maximum load factor, limited by stall or buffet ($w = 4\,\text{kN/m}^2$).

$$C_{L0} = 2w/(\rho_0 \sigma V^2),$$

in which $\sigma = 0.6009$ for $h = 5\,\text{km}$.

(a) For $V = 100\,\text{m/s}$ the lift coefficient in straight flight is

$$C_{L0} = 0.951.$$

Then Equation (8.9) gives the maximum load factor as

$$n = C_L/C_{L0} = 1.2/0.951 = 1.262,$$

from which Equation (8.7) gives $\phi = 37.6°$ and Equation (8.6) gives $V^2/(gR) = 0.770$. Hence the turn radius is $R = 1.324\,\text{km}$ and the turn rate is $\Omega = V/R = 0.0755\,\text{rad/s}$. The time for one complete turn must then be $2\pi/\Omega = 83.2$ seconds.

(b) Similarly, for $V = 200\,\text{m/s}$, $C_{L0} = 0.238$. Then the maximum n is $0.8/0.238 = 3.361$, $\phi = 72.7°$ and $V^2/(gR) = 3.209$, from which $R = 1.271\,\text{km}$ and $\Omega = V/R = 0.1574\,\text{rad/s}$. One complete turn will then take $2\pi/\Omega = 39.9$ seconds.

228   Turning performance

When the speed is doubled there is a large increase in maximum n and ϕ but in this particular case there is only a small change in the minimum radius.

8.5   Rate of climb in a banked turn

As mentioned earlier there is an increase of drag when an aircraft changes at constant speed from straight flight to a banked turn and this reduces the rate of climb. An important special case occurs when the maximum rate of climb in the turn is zero at a specified speed; the rate of turn and the load factor then have the maximum values that can be sustained at that speed without loss of height.

It will be assumed that in the turn the angle of climb γ is small enough to justify the assumption that $\cos \gamma = 1$. The flight path is then a close-coiled helix and all the Equations (8.4)–(8.16) are valid without change. In general the aircraft may be accelerating along the flight path and, as explained in Chapter 4, allowance can be made for this by deriving equations for V_c', the rate of increase of the energy height h_e. Equation (4.32) shows that

$$V_c' = dh_e/dt = P_{se} = V(F - D)/W \qquad (8.17)$$

and this equation can be used to show how V_c' decreases as the drag rises in a change from straight flight to a banked turn. It will be assumed that any increase of drag due to the angular velocity of the aircraft and the deflections of the control surfaces can be neglected in comparison with the dominant increase of lift-dependent drag. With no change of speed or height the thrust F remains constant and the reduction of P_{se} (or V_c') is

$$\Delta P_{se} = V \, \Delta D / W, \qquad (8.18)$$

where ΔD is the increase of drag. In general, if data are available giving C_D as a function of C_L for the relevant Mach number, the values of ΔD and ΔP_{se} can be found from these data. If the simple parabolic drag law can be assumed to be valid Equation (3.6) shows that

$$\Delta D = \tfrac{1}{2}\rho V^2 S \, \Delta C_D = \tfrac{1}{2}\rho V^2 S K_2 \, \Delta(C_L^2), \qquad (8.19)$$

where $\Delta(C_L^2)$ is the increase of C_L^2, given by

$$\Delta(C_L^2) = (n^2 - 1) C_{L0}^2 \qquad (8.20)$$

and C_{L0} is the lift coefficient in straight flight at the same speed V.

Since

$$W = \tfrac{1}{2}\rho V^2 S C_{L0}, \qquad (8.21)$$

Rate of climb in a banked turn

Equations (8.18)–(8.20) show that

$$\Delta P_{se} = VK_2 C_{L0}(n^2 - 1) \tag{8.22}$$

or alternatively

$$\Delta P_{se} = 2wK_2(n^2 - 1)/(\rho V). \tag{8.23}$$

Figure 8.10 shows ΔP_{se} as given by Equation (8.23), plotted against Mach number and speed for an aircraft with $w = 4 \text{ kN/m}^2$ and $K_2 = 0.075$ at a height of 5 km. This figure also shows the boundaries for stall margin and buffet represented by lift coefficients of 1.0 and 0.75, as discussed earlier in connection with Figures 8.4–8.9. It should be emphasised that Figure 8.10 is based on the simple parabolic drag law with a constant value of the coefficient K_2, whereas in reality K_2 usually increases at high Mach numbers, especially for higher values of

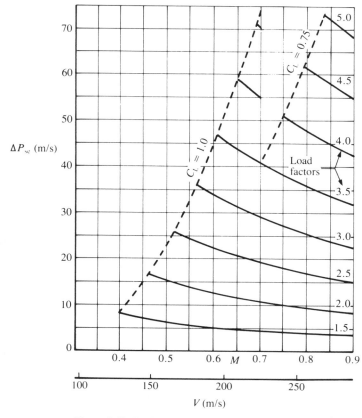

Figure 8.10. Reduction of P_{se} due to turn. $w = 4 \text{ kN/m}^2$, $K_2 = 0.075$, $h = 5$ km.

230   *Turning performance*

C_L. Thus if $K_2 = 0.075$ at moderate values of C_L and low Mach numbers, ΔP_{se} may be greater than is shown in the figure for the higher values of M because K_2 is then likely to be greater than 0.075.

The curves for constant n in Figure 8.10 are rectangular hyperbolae because Equation (8.23) shows that for constant w, K_2, ρ and n the reduction of P_{se} is inversely proportional to the speed. The same equation also shows that for a given speed $\Delta P_{se} \propto (n^2 - 1)$.

In the general case where the flight speed is not constant the reduction of P_{se} shown in Figure 8.10 is equal to the reduction of V_c' at a given instantaneous speed and Equation (4.32) shows that this implies a reduction in either the rate of climb V_c or the forward acceleration, or both of these. If the flight speed is constant there is no component of acceleration along the flight path and $P_{se} = V_c$, so that the reduction of P_{se} is equal to the reduction in the rate of climb. In this case the angle of climb is given by $\sin \gamma = P_{se}/V$ and if γ is small the reduction of angle of climb in changing from straight to turning flight is

$$\Delta \gamma = \Delta P_{se}/V. \tag{8.24}$$

An alternative approach to the estimation of the rate of climb in a banked turn is based directly on the equations for turning flight, rather than a comparison with straight flight. Since $C_L = nC_{L0}$ and $C_{L0} = C_L^*/v^2$, the lift coefficient in the turn can be expressed as $C_L = nC_L^*/v^2$ and then the appropriate drag can be introduced by using Equation (3.11) to obtain

$$\beta = K_1 v^2/(nC_L^*) + K_2 nC_L^*/v^2. \tag{8.25}$$

Equations (3.12) and (3.13), which are still valid in a turn, then lead to the more useful form

$$\beta = \tfrac{1}{2}\beta_m(v^2/n + n/v^2). \tag{8.26}$$

Alternatively, since $n/v^2 = C_L/C_L^*$,

$$\beta = \tfrac{1}{2}\beta_m(C_L/C_L^* + C_L^*/C_L) \tag{8.27}$$

and this is the same as Equation (3.19), noting that n_1 in that equation is defined as C_L/C_L^*. In turning flight Equation (8.26) replaces Equation (3.18), which is valid only for straight flight with $n = 1$. Note also that in turning flight the drag is

$$D = (L/W)(D/L)W = n\beta W \tag{8.28}$$

and for flight at constant speed Equations (4.6) and (4.7) are replaced by

$$V_c = (V_e/\sigma^{1/2})(f - n\beta) \tag{8.29}$$

Rate of climb in a banked turn

and

$$V_c = [V_e^* \beta_m / \sigma^{1/2}][v\epsilon - \tfrac{1}{2}(v^3 + n^2 v^{-1})]. \tag{8.30}$$

The angle of climb is then given by

$$\sin \gamma = V_c/V = \beta_m [\epsilon - \tfrac{1}{2}(v^2 + n^2 v^{-2})]. \tag{8.31}$$

As explained in Chapter 4 the rate of climb V_c is equal to P_{se} when the speed is constant and in the more general case where the speed is not constant V_c in Equations (8.29) and (8.30) can be replaced by the rate of increase of energy height V_c', which is again equal to P_{se}. Then Equation (8.30) can be used to find the value of P_{se} for any specified values of n, v, ϵ, β_m and V^* ($= V_e^*/\sigma^{1/2}$). It can also be used to find the loss of P_{se} in a change from straight flight (with $n = 1$) to turning flight with a specified value of n. The equation shows that this loss of P_{se} is

$$\Delta P_{se} = \tfrac{1}{2} V_e^* \beta_m (n^2 - 1)/(v\sigma^{1/2}) \tag{8.32}$$

and substitution here of Equations (3.13) and (3.17) for β_m and V_e^* shows that this is the same as Equation (8.23).

Using Equation (8.30) with V_c replaced by P_{se}, Figure 8.11 shows P_{se} plotted against Mach number and speed for an aircraft with $w = 4 \text{ kN/m}^2$, $K_1 = 0.025$, $K_2 = 0.075$ and $f = 0.5$ at a height of 5 km. These values are the same as those used in Figure 8.10 but in addition it has been necessary for evaluation of P_{se} to assume values for K_1 and f. For the values given here $\beta_m = 0.0866$, $V_e^* = 106.35 \text{ m/s}$ and $\epsilon = 5.77$. For any specified speed the difference of P_{se} between $n = 1$ and any larger value of n is of course the value given by Figure 8.10. It should be noted again that the results given are based on the simple parabolic drag law, with constant values of K_1 and K_2 and they may overestimate P_{se} at high Mach numbers, especially at high values of n.

The broken lines in Figure 8.11 show the conditions for constant lift coefficients of 1.0 and 0.75, representing the assumed stalling and buffeting limits discussed earlier. These show that, for the assumed drag law and for this large thrust/weight ratio of 0.5, a positive P_{se} can be maintained for any load factor that is allowed by the stalling and buffeting limits up to a Mach number of at least 0.85. With a lower value of the thrust/weight ratio f all the values of P_{se} would be reduced and for the higher load factors P_{se} might be expected to fall below zero, indicating that a 'high-g' turn could be maintained only with deceleration or loss of height, or both of these.

EXAMPLE 8.2. DECELERATION IN A TURN DUE TO THRUST DEFICIT

An aircraft has a wing loading of 4 kN/m^2, with $K_1 = 0.025$ and $K_2 = 0.075$. Find the value of f needed to give a rate of climb of

232 *Turning performance*

50 m/s in straight flight at a height of 5 km and a constant speed of 200 m/s. With this value of f the aircraft makes a banked turn with $n = 2.5$, starting at the same speed and height. If the rate of climb in the turn is maintained at 40 m/s find the rate of loss of flight speed.

As for the aircraft represented in Figure 8.11, $\beta_m = 0.0866$, $V_e^* = 106.35$ m/s and $V^* = V_e^*/\sigma^{1/2} = 137.2$ m/s. Hence $v = V/V^* = 1.458$ and for straight flight Equation (3.18) gives $\beta = 0.1124$. Then Equation (4.6) or (8.29) shows that for the required rate of climb

$$f = V_c/V + \beta = 50/200 + 0.1124 = 0.3624.$$

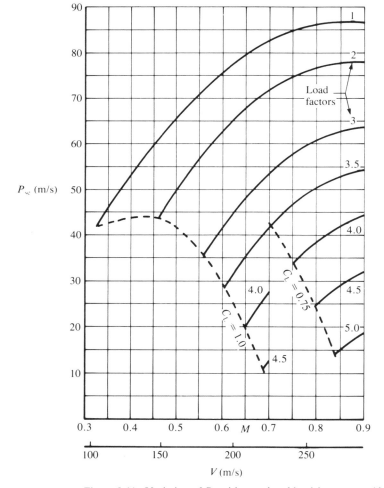

Figure 8.11. Variation of P_{sc} with speed and load factor. $w = 4 \text{ kN/m}^2$, $K_1 = 0.025$, $K_2 = 0.075$, $f = 0.5$, $h = 5$ km.

For a turn with $n = 2.5$, Equation (8.23) gives

$$\Delta P_{se} = 21.40 \text{ m/s}$$

and hence

$$P_{se} = 50 - 21.4 = 28.6 \text{ m/s}.$$

(This value of P_{se} can also be obtained from Equation (8.30), with $\epsilon = f/\beta_m = 4.185$.) Then Equation (4.34) shows that with $V_c = 40$ m/s the rate of change of flight speed is

$$dV/dt = (g/V)(P_{se} - V_c) = -0.559 \text{ m/s}^2.$$

8.6 The thrust boundary for a banked turn at constant height

It has been shown that for a given thrust there is a substantial reduction of P_{se} as the load factor in a turn increases. Thus if an aircraft is required to turn without any loss of either height or speed the maximum allowable load factor may be limited by the available thrust.

With V_c replaced by P_{se} Equation (8.30) shows that the condition for $P_{se} = 0$ is

$$\epsilon = \tfrac{1}{2}(v^2 + n^2 v^{-2}) \qquad (8.33)$$

and since $P_{se} = V(F - D)/W$ this is also the condition for $F = D$. The lines of constant load factor in Figure 8.12 are derived from this equation and show the minimum values of ϵ required to maintain $P_{se} \geq 0$. Thus if there is a requirement that the aircraft should maintain height in a turn without loss of speed the ratio $\epsilon = f/\beta_m$ must not be less than the value given by Equation (8.33) or Figure 8.12. It should be noted that the curves in the figure apply universally to any aircraft at any height provided the simple parabolic drag law is valid, but they may underestimate the required values of ϵ at large v and high altitude, where the Mach number is high and the drag may be increased.

The speed ratio v is equal to $V_e/V_e^* = (C_L^*/C_{L0})^{1/2}$, where C_{L0} is the lift coefficient at the EAS V_e in straight flight. In a turn at the same EAS the lift coefficient is $C_L = nC_{L0}$ and hence

$$v = (nC_L^*/C_L)^{1/2}. \qquad (8.34)$$

Equation (8.33) shows that for a given load factor the required value of ϵ is minimum when $v = n^{1/2}$ and ϵ is then equal to n. Equation (8.34) shows that for this condition $C_L = C_L^*$, so that $\beta = \beta_m$. These features are illustrated in Figure 8.12 where the line $C_L = C_L^*$ can be seen to pass through the minima of all the lines of constant n, crossing each at $v = n^{1/2}$. This line $C_L = C_L^*$, like the lines of constant n, applies to any aircraft at any height. If ϵ is independent of speed for a

234 *Turning performance*

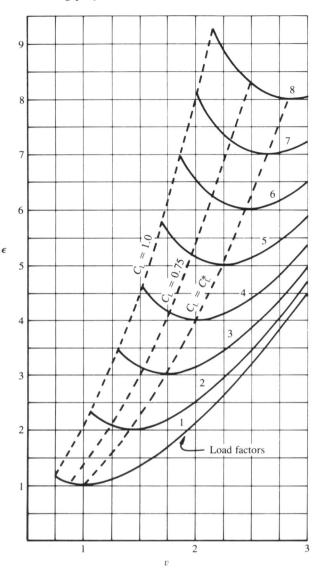

Figure 8.12. Thrust ratio $\epsilon = f/\beta_m$ required for turn without loss of height or speed. Lines of constant C_L are calculated for $C_L^* = (\frac{1}{3})^{1/2}$.

given height, the line $C_L = C_L^*$ also gives the condition for the ceiling of the aircraft expressed in terms of the load factor in a turn. Thus, for example, the '3g ceiling' is the greatest height at which a thrust of $3\beta_m W$ is available, since this gives $\epsilon = 3$ and allows a turn with $n = 3$ (and $v = 3^{1/2}$) without loss of height or speed. Figure 8.12 shows that it is never possible to maintain height and speed in a turn with a load factor n that is greater than the maximum available value of ϵ.

Figure 8.12 also shows curves for the constant values of C_L, 1.0 and 0.75, representing the stall and buffet boundaries considered earlier. These curves have been derived from Equation (8.34) assuming that $C_L^* = (\frac{1}{3})^{1/2}$, a value that is consistent with the values of K_1 and K_2 assumed in deriving Figure 8.11 (because $C_L^* = (K_1/K_2)^{1/2}$). The curves apply to any aircraft at any height, provided that C_L^* has this value.

As a further illustration of the effects of thrust limitations on the rate of turn when no loss of height or speed can be allowed, some lines of constant ϵ are drawn on Figures 8.7 and 8.8. These are calculated for $P_{se} = 0$ by using Equations (8.33) and (8.16), assuming that $w = 4 \text{ kN/m}^2$ and taking C_L^* to be $(\frac{1}{3})^{1/2}$, as for the lines of constant C_L in Figure 8.12. These lines in Figures 8.7 and 8.8 show several of the features considered earlier in discussion of Figure 8.12; each line of constant ϵ touches a line of constant n having the value $n = \epsilon$ and the point of contact represents the condition $v = n^{1/2}$, where ϵ is minimum for a given n or n is maximum for a given ϵ.

The lines of constant ϵ in Figures 8.7 and 8.8 show the boundaries within which V and n must lie, for any specified level of thrust, if there is to be no loss of speed or height. For instance, if the thrust available at sea level is sufficient to allow $\epsilon = 4$ up to $M = 0.8$, Figure 8.7 shows that $n = 3$ is obtainable up to that Mach number, but at $M = 0.85$ the maximum value of n is only 2, even if $\epsilon = 4$ is still available at the higher speed. A turn can be maintained without loss of speed or height only if the speed and load factor are such that the point representing these conditions on a diagram such as Figure 8.7 or 8.8 lies below the line of constant ϵ for the value of ϵ that is available at the relevant speed.

Figures 8.7 and 8.8 show that for a given value of ϵ the maximum rate of turn Ω is obtained at a speed that is less than the value giving maximum n. Equations (8.33) and (8.16) show that for $P_{se} = 0$ at any given height

$$\Omega \propto (2\epsilon - v^2 - v^{-2})^{1/2}$$

and for a given value of ϵ this is maximum when $v = 1$. Since C_{L0} is then equal to C_L^*, Equation (8.9) shows that $C_L = nC_L^*$ when Ω is maximum. For the aircraft considered in Figures 8.7 and 8.8 the corresponding Mach numbers are 0.313 at sea level and 0.612 at $h = 10$ km. The maxima can be seen on both the lines of constant ϵ in Figure 8.8 and for $\epsilon = 2$ in Figure 8.7, but for the higher values of ϵ shown in Figure 8.7 the maxima occur at lift coefficients above the assumed limit of 1.0 and are not shown.

EXAMPLE 8.3. MAXIMUM VALUES OF ANGLE OF BANK AND RATE OF TURN, FOR NO LOSS OF SPEED OR HEIGHT

An aircraft has a wing loading of $4\,\text{kN/m}^2$, with $K_1 = 0.025$, $K_2 = 0.075$ and $f = F/W = 0.17$. If there is to be no loss of speed or height in a banked turn at a height of 5 km, find the maximum values of (a) the angle of bank and (b) the rate of turn. Also find the lift coefficient, the true air speed and the radius of turn for the flight conditions giving the maxima (a) and (b).

As in Example 8.2, $\beta_m = 0.0866$ and $V^* = 137.2$ m/s. For maximum angle of bank n is maximum and hence

$$n = \epsilon = f/\beta_m = 1.963 \quad \text{and} \quad \phi = \sec^{-1} n = 59.4°.$$

Then $v = n^{1/2} = 1.401$ and $V = vV^* = 192.2$ m/s. Equation (8.34) shows that

$$C_L = C_L^* = (K_1/K_2)^{1/2} = 0.5774$$

and the radius of turn is given by Equation (8.12) as 2.229 km.

For maximum rate of turn $v = 1$ and $V = V^* = 137.2$ m/s. Then Equation (8.33) shows that

$$n = (2\epsilon - 1)^{1/2} = 1.7106$$

and Equation (8.16) gives the rate of turn as

$$\Omega = 0.0992 \text{ rad/s}.$$

The lift coefficient is $nC_L^* = 0.9877$ and the radius of turn is given by Equation (8.12) as 1.383 km.

9

Vectored thrust

An aircraft with vectored thrust is defined here as one in which the pilot is able to vary the *direction* of the engine thrust over a wide range, usually at least 90°. The main advantage of this facility is that if the maximum available thrust is greater than the weight, the aircraft is able to take-off and land vertically, i.e. with zero ground run. If the thrust is large but still less than the take-off weight it may be possible to use thrust vectoring to give a substantial reduction in the distance required for take-off and in this case the landing weight may be less than the available thrust so that a vertical landing may be possible. Aircraft with vectored thrust are commonly known as V/STOL aircraft because they are capable of vertical or short take-off and landing.

V/STOL aircraft have been designed with several distinct configurations, the best known being that used in the Harrier as shown in Figure 9.1 and described by Fozard (1986), where the propulsive jets can be deflected downward by movable nozzles. Other forms have been reviewed by Poisson-Quinton (1968) and include the tilt rotor, where lifting rotors of the kind used in helicopters are tilted forward to operate like normal propellers in forward flight, and the tilt wing where the rotor axes remain fixed in the wing and the whole wing–rotor assembly is rotated relative to the fuselage. An example of a tilt-rotor aircraft is the Bell Textron Osprey shown in Figure 9.2. This chapter is written mainly with reference to the Harrier type configuration, because that method of obtaining a wide variation in thrust direction is more versatile than some others such as the tilt rotor and because there has been more operational experience with it than with the others, but many of the equations and conclusions are approximately valid for any form of aircraft in which the thrust direction is continuously variable.

For the full exploitation of the advantages of vectored thrust some unsteady motions involving transient conditions are needed but

238   *Vectored thrust*

Figure 9.1. The British Aerospace Sea Harrier in hovering flight. The downwardly deflected hot jets can be seen as two regions of hazy appearance below the aircraft.

Figure 9.2. The Bell Textron Osprey tilt-rotor aircraft in transition from hovering flight, supported by the rotors, to forward flight supported by the wing.

attention will be concentrated here on some steady-state conditions, making a number of simplifying assumptions. The use of vectored thrust for vertical or short take-off and landing and for turning flight will also be reviewed briefly.

In general there is substantial interference between the deflected jets and the aerodynamic lift and there are also important effects of ground proximity, usually tending to reduce the overall upward vertical force. All these effects will be neglected in the analysis given here and another important simplifying assumption will be made concerning the intake momentum drag of a turbofan engine, mentioned in § 5.2. For a conventional turbofan aircraft without vectored thrust this term can be properly regarded as a loss of thrust as given by the term $\dot{m}V$ in Equation (5.1), and as explained in § 5.2 only the *net* thrust of the engine needs to be considered in performance calculations. With vectored thrust of the kind used in the Harrier, the intake momentum drag should be regarded as an addition to the drag of the aircraft because it acts along the line of the flight path, even when the thrusting jets are deflected, and the deflected thrust used in performance analysis should be the *gross* thrust. Nevertheless in order to derive equations and results which are approximately valid for all forms of vectored thrust aircraft the distinction between net and gross thrust will be ignored here, noting that the difference between these two thrusts is the intake momentum drag and is small at low flight speeds. The thrust F and the drag D will both be assumed to be independent of the inclination of the thrust line to the flight direction.

It has been assumed in all the preceding chapters that the thrust F acts along the flight path. The derivation of equations for flight with inclined thrust in the present chapter provides a convenient basis for assessing the error introduced by this assumption for conventional cruising and climbing flight, and for estimating the optimum setting of the thrust line for economy in the cruise or climb.

9.1 Equations for steady flight

Figure 9.3 shows the forces acting on an aircraft with vectored thrust. The notation is the same as that shown in Figure 2.1 except that the thrust vector is now inclined upward at an angle θ_F from the flight path. It will be recognised that the pilot actually sets the direction of the thrust vector relative to an aircraft datum line, by adjustment of the nozzles. Using θ to denote the inclination of the aircraft datum to the horizontal, the angle of incidence referred to this datum is $\theta - \gamma$ and this angle varies widely with the condition of flight, being positive in normal wing-borne flight but negative for the initial period after a vertical take-off, as shown in Figure 9.3. In the latter state the aerodynamic lift would have a negative value if the datum

240 Vectored thrust

line were chosen to be in the direction for which a coincident flight path would give zero lift. Thus the angle θ_F is not simply related to the nozzle angle as set by the pilot, but for any specified condition of steady flight a nozzle angle can be found which gives the required value of θ_F.

For steady flight with zero acceleration the forces are in equilibrium and must satisfy the equations

$$L + F \sin \theta_F = W \cos \gamma \tag{9.1}$$

and

$$F \cos \theta_F = D + W \sin \gamma. \tag{9.2}$$

Assuming that $\cos \gamma = 1$, introduction of the ratios $n = L/W$, $f = F/W$ and $\beta = D/L$ then gives

$$f \sin \theta_F = 1 - n \tag{9.3}$$

and

$$f \cos \theta_F = n\beta + \sin \gamma, \tag{9.4}$$

leading to

$$\tan \theta_F = (1 - n)/(n\beta + \sin \gamma). \tag{9.5}$$

Also, the Equations (8.26) and (8.27) give β correctly for this case (with $n < 1$), even though they were derived with reference to turning flight (with $n > 1$), since the derivation does not depend in any way on the mechanism causing n to differ from 1. Equation (8.26) shows that the $n\beta$ term in Equations (9.4) and (9.5) is

$$n\beta = D/W = \tfrac{1}{2}\beta_m(v^2 + n^2 v^{-2}) \tag{9.6}$$

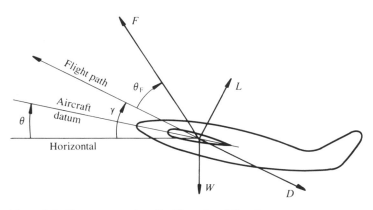

Figure 9.3. Forces on an aircraft with vectored thrust.

Level flight at low speed 241

and this may be used with Equation (9.5) to give θ_F in terms of β_m, n, v and γ.

9.2 Optimum values of θ_F for cruise and climb

Equation (9.3) shows that for constant f the lift ratio n decreases as the angle θ_F increases from zero. Equations (9.6) and (9.4) then show that for given values of the speed and angle of climb, $n\beta$ and $f \cos \theta_F$ also decrease and it is of interest to consider the value of θ_F requiring minimum f.

Assuming that $\beta_m = 0.08$, calculations of β, θ_F and f have been made for a range of values of n, with $v = 1.0$ and 2.0 and for $\gamma = 0$, $10°$ and $20°$. For each case there is a value of n, and hence of θ_F, for which f has the minimum value f_{min} and Figure 9.4 shows the ratio f/f_{min} plotted against θ_F for various values of γ and v. Further calculations have shown that the effects of varying β_m are relatively small; an increase by 50% to $\beta_m = 0.12$, for example, reduces f/f_{min} by amounts up to only 0.2% for $v = 2$ and only 0.4% for $v = 1$. Similarly, a decrease to $\beta_m = 0.055$ increases f/f_{min} by comparable amounts. The value of θ_F for minimum f is increased by less than $2°$ when β_m is increased to 0.12 and reduced by less than $1\frac{1}{2}°$ when β_m is reduced to 0.055.

As shown in Figure 9.4 there is an increase in the value of θ_F for minimum f when C_L rises, i.e. when v decreases. For turbofan aircraft the value of v in climbing and cruising flight is almost always greater than 1 and hence for $\beta_m = 0.08$ the optimum value of θ_F will be no more than about $5°$, rising only to about $7°$ if β_m is increased to the exceptionally high value of 0.12. With the thrust line fixed in the aircraft, θ_F is equal to $(\alpha + \text{constant})$, where α is the angle of incidence, so that θ_F increases with C_L and there should be no difficulty in designing the aircraft so that in cruising and climbing flight θ_F never differs from the optimum value by more than about $5°$, requiring a thrust that is no more than 0.5% above the minimum.

The assumption has been made in all the earlier chapters that the thrust always acts along the flight path, i.e., that $\theta_F = 0$. Figure 9.4 shows that this is a good approximation if the true value of θ_F does not exceed about $10°$, since the error in the required thrust is then no more than about 1.2%, for values of v up to 2.0 and for climb angles up to $20°$.

9.3 Level flight at low speed

A V/STOL aircraft with $F \geq W$ can remain airborne at zero forward speed, but even if F is somewhat less than W the aircraft can remain aloft by using the vertical component of thrust to support a large part of the weight, while flying slowly forward to satisfy the much reduced requirement for aerodynamic lift. (It is then possible by flying in a circle of quite small radius to achieve conditions near to hover.) This mode of flight is sometimes known as partially jet-borne (PJB) flight. For flight at the lowest possible speed the aerodynamic lift coefficient should be as high as possible and it is convenient to express the maximum usable C_L as a multiple of the lift coefficient C_L^* giving minimum β. As noted in § 9.1, Equation (8.27) is valid for this case and gives β in terms of β_m and C_L/C_L^*. If v_1 is used to denote the value of v at a specified value of C_L for $n=1$ the ratio v/v_1 gives a measure of the fraction of speed required at constant C_L when

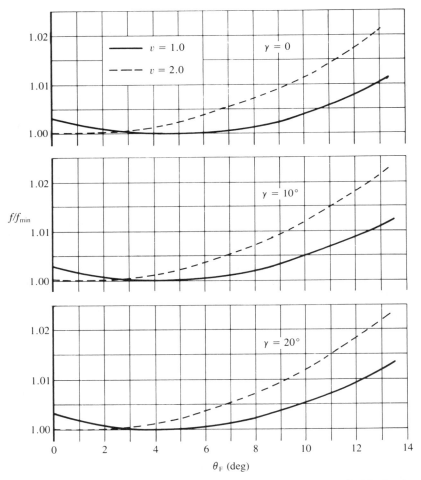

Figure 9.4. Effect of varying thrust direction on the thrust required for level flight and for climb ($\beta_m = 0.08$).

Level flight at low speed

vectored thrust is available. Equation (8.34) is valid for this case and shows that

$$n/v^2 = 1/v_1^2$$

and hence

$$v/v_1 = n^{1/2}. \tag{9.7}$$

For level flight at any selected value of v/v_1, the ratio n is given by Equation (9.7) and with $\gamma = 0$ Equations (9.5) and (9.3) give the required values of θ_F and f.

The full lines in Figure 9.5 show results obtained from these equations for $\beta_m = 0.08$ and $C_L/C_L^* = 2$. For these values, $v_1 = 2^{-1/2} = 0.707$ and Equation (8.27) gives $\beta = 0.10$. With $f = 1$, flight at zero speed is just possible with $\theta_F = 90°$. At the other extreme, with $v = v_1$ and $\theta_F = 0$, Equation (9.4) shows that $f = \beta$. There is a useful range of intermediate conditions, e.g. $f = 0.75$ gives $v/v_1 = 0.5$ with $\theta_F = 88°$ and $f = 0.51$ gives $v/v_1 = 0.7$ with $\theta_F = 84.5°$. The required value of θ_F exceeds 80° for all values of f greater than about 0.35 ($v/v_1 < 0.8$) and within this range $\sin \theta_F \approx 1$ and Equation (9.3) shows that

$$f \approx 1 - n. \tag{9.8}$$

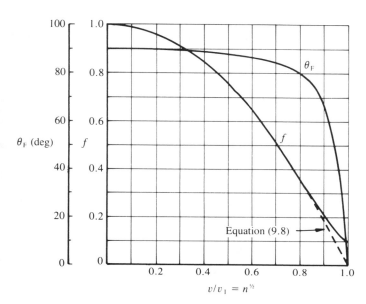

Figure 9.5. Speed in partially jet-borne level flight with $\beta_m = 0.08$ and $C_L/C_L^* = 2.0$.

244 Vectored thrust

The broken line in Figure 9.5 represents this equation and confirms that it is almost exact for $v/v_1 < 0.8$, i.e. for $n < 0.64$.

For fuel economy, it is clear that f should be reduced as far as possible and Figure 9.5 shows that the required value of f drops quickly as speed builds up from the hover, but the thrust vector stays remarkably vertical for quite substantial values of the speed fraction v/v_1. Alternatively, because v_1 is a limiting speed for conventional flight at the maximum usable C_L, it is clear from Figure 9.5 that as the speed drops from v_1 to fractional values of v/v_1, the thrust inclination θ_F must rise rapidly to allow for loss of aerodynamic lift. At the same time the additional thrust beyond what is necessary for conventional level flight ($f = 0.1$) will also rise rapidly, doubling for example when the forward speed has only dropped to 90% of the limiting speed v_1.

EXAMPLE 9.1. THRUST CONDITIONS REQUIRED FOR SPECIFIED REDUCTION OF MINIMUM SPEED

An aircraft has a wing loading of 4 kN/m² and a drag coefficient given by Equation (3.6) with $K_1 = 0.025$ and $K_2 = 0.075$. In level flight at sea level the minimum safe speed is 70 m/s when there is no vertical component of thrust. If this speed is to be reduced to 55 m/s by the use of vectored thrust, without any change of lift coefficient, find the minimum value of the ratio $F/W = f$ and the required inclination of the thrust line to the horizontal.

For the determination of β, preliminary calculations are needed to find β_m and C_L/C_L^*.

$$\beta_m = 2(K_1 K_2)^{1/2} = 0.0866 \text{ and } C_L^* = (K_1/K_2)^{1/2} = 0.5773.$$

With no vertical thrust component

$$C_L = w/(\tfrac{1}{2}\rho_0 V_e^2)$$
$$= 8000/(1.225 \times 70^2) = 1.333.$$

Hence $C_L/C_L^* = 2.308$ and Equation (8.27) gives $\beta = 0.1187$. The required speed ratio v/v_1 is 55/70 and therefore $n = (v/v_1)^2 = 0.6173$. Then Equation (9.5) gives $\theta_F = 79.16$ and Equation (9.3) gives $f = 03897$. (Note that in this case $\sin \theta_F \approx 1$ and f is close to $(1-n) = 0.3827$.)

9.4 Vertical take-off and landing

If f can be greater than 1 the aircraft can lift off vertically with the thrust line vertical, and it is usually found that the thrust required for this is rather greater than that needed for hover in free air, because in most cases there is an adverse effect of ground proximity. With the aircraft hovering well above the ground the transition to wing-borne forward flight is started by reducing the inclination of the thrust line to produce a forward horizontal component of thrust. Fozard (1986) has reported that with the Harrier the forward acceleration during this transition approaches 0.5 g and the total time to change from $n = 0$ to $n = 1$ can be less than 30 seconds. As the drag is small in PJB flight at low speeds, it can be shown that Equations (9.1) and (9.2) predict the acceleration given by Fozard for a thrust of approximately 1.15 W oriented at $\theta_F \approx 60°$.

Fozard (1986) has also given an account of the procedure used for vertical landing of the Harrier. The decelerating transition from wing-borne to jet-borne flight is normally completed in less than 30 seconds, implying that the deceleration again approaches 0.5 g. The direction of flight during transition can be independent of wind direction and the aim is to end the transition about 15 or 20 m above the ground within sight of the required landing spot, using the final hovering phase to remove any errors of position or heading.

9.5 Short take-off

When the maximum quantities of fuel and payload are carried in the aircraft the total weight is often greater than the maximum installed thrust. Take-off can then be achieved only by augmenting the vertical component of thrust with aerodynamic lift, using a ground run to build up the forward speed. For almost the whole of the ground run the angle θ_F is kept at a value close to zero and with full thrust the acceleration approaches 1 g because f is usually not much less than 1. At a predetermined speed V_R the thrust line is rotated quickly and held at the new setting until the aircraft has cleared the hypothetical 'screen' as explained in Chapter 6. When f is relatively large (although less than 1) the aircraft usually lifts off soon after rotation of the thrust line, without any deliberate rotation of the aircraft, but for smaller values of f the pilot may need to rotate the aircraft by use of the elevator control. In any case the speeds for rotation of the thrust line and for rotation of the aircraft (if needed) are almost identical and are very close to the lift-off speed V_{LO} and this justifies the use of the symbol V_R here for the thrust rotation speed, although in Chapter 6 it was defined as the *aircraft* rotation speed.

As suggested by Taylor (1975), both the length of ground run and the distance to the screen height can be greatly reduced by using an upwardly inclined ramp for the take-off. The complete ramp structure is concave, so that the aircraft starts its ground run horizontally and becomes airborne as it leaves the ramp, with its flight path then inclined upward from the horizontal at an angle that is typically in the range 10° to 15°. Taylor (1975) introduced the name 'ski-jump' for the ramp and the take-off technique, because the ramp is used like a ski-jump in reverse. Immediately after the aircraft leaves the ramp the flight path curves downward briefly because the total effective lift, including the thrust component $F \sin \theta_F$, is still less than the weight. As the speed increases, the aerodynamic lift builds up and the flight path first straightens and then starts to curve upward. The ski-jump has been much used for launching V/STOL aircraft from ships as in Figure 9.6, but in the following discussion it will be assumed that the aircraft takes off from a conventional flat runway or deck, not using a ski-jump.

The calculation of short take-off performance for V/STOL aircraft of the Harrier class is explained in detail in ESDU 87037 (1988), both for a flat runway and for a ski-jump. For the former, it is emphasised that the two most important variables controlling the take-off performance are the thrust rotation speed V_R and the thrust inclination θ_F after rotation. It is clear that the aircraft cannot lift off unless V_R exceeds the minimum value for PJB flight with an acceptable value of C_L, as discussed in § 9.3. There is a choice of V_R above this minimum value and information given in ESDU 87037 (1988) is represented in

Figure 9.6. Sea Harrier taking off from a ship fitted with a ski-jump ramp. The aircraft has just left the ramp and might lose some height during the next few seconds.

Short take-off

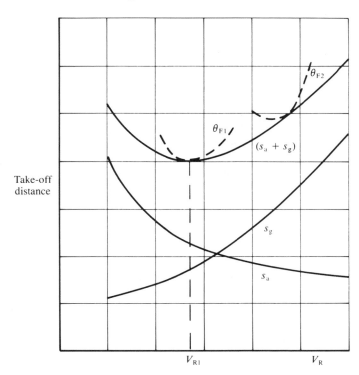

Figure 9.7. Variation of take-off distance with thrust rotation speed and thrust angle.

Figure 9.7 and shows in qualitative terms how the take-off distance varies with V_R and θ_F. As shown in Figure 6.6 the total take-off distance is made up of two parts, a ground distance s_g and an airborne distance s_a. The two curves shown by broken lines in Figure 9.7 show how the total distance $(s_a + s_g)$ varies with V_R for two fixed values of the thrust angle θ_{F1} and θ_{F2} with $\theta_{F1} > \theta_{F2}$. For each value of θ_F there is an optimum value of V_R giving minimum distance and the curve marked $(s_a + s_g)$ and shown by a full line is the envelope of all the curves drawn for constant θ_F. This envelope has a minimum at $V_R = V_{R1}$ and the thrust angle required for this is θ_{F1}. The two curves marked s_a and s_g show how the envelope curve for $(s_a + s_g)$ is made up of the two components s_a and s_g. If V_R is reduced below V_{R1} while keeping θ_F at the optimum value for each speed, the increase of s_a exceeds the reduction of s_g, whereas if V_R is increased above V_{R1} the increase of s_g is greater than the reduction of s_a. It has been reported by Fozard (1986) that the optimum value of V_R for the Harrier usually gives a value of n that is no more than about 0.25–0.3 at lift-off, i.e. less than one third of the effective lift is then of aerodynamic origin.

248   Vectored thrust

For any given value of V_R and for $\theta_F \approx 0$ during the ground run the distance s_g covered in the ground run can be calculated as for a conventional take-off, using one of the methods described in § 6.7. The calculation of the airborne distance s_a from lift-off to the screen requires a step-by-step integration of the equations of motion. The equations are essentially those already given for PJB flight, but with the addition of terms to represent the components of acceleration along and normal to the flight path, although ESDU 87037 (1988) states that the acceleration along the flight path is usually small in the airborne phase of the take-off. During this phase the angle of incidence, and hence the lift coefficient, is limited by the need for safe handling of the aircraft and is usually kept nearly constant.

9.6   The use of vectored thrust in a turn

A high rate of turn is a desirable feature of a combat aircraft. When the maximum turn rate is determined by human limitations as discussed in § 8.3 no improvement can be obtained by the use of vectored thrust, but when the turn is limited either by wing aerodynamics (stalling or buffeting) or by the structural design, the rate of turn can be increased by the use of vectored thrust to augment the total force in the direction of the lift. Putting $L' = F \sin \theta_F$, the component of thrust acting in the lift direction, Equations (8.4) and (8.5) for a true-banked turn in a horizontal plane can be replaced by

$$mg = (L + L') \cos \phi \tag{9.9}$$

and

$$mV^2/R = (L + L') \sin \phi. \tag{9.10}$$

Equation (8.6) is unchanged and if

$$n' = L'/W = f \sin \theta_F,$$

Equation (8.7) is replaced by

$$n + n' = (L + L')/W = \sec \phi$$
$$= (1 + \tan^2 \phi)^{1/2} = [1 + V^4/(gR)^2]^{1/2}. \tag{9.11}$$

The lift coefficient C_L is proportional to n, not $(n + n')$, and Equation (8.9) is unchanged. Hence for any specified maximum value of C_L and for given wing loading and height the minimum radius of turn can be found as a function of speed if a value is assumed for $n' = f \sin \theta_F$. As an example, Figure 9.8 shows how the minimum radius of turn can be reduced by the use of vectored thrust, for an aircraft with $w = 4 \text{ kN/m}^2$ and with C_L limited to 1.0. The full lines show the radius for $n' = 0$, i.e. for $\theta_F = 0$, at the two heights of 0 and 10 km and these are the same as the broken lines shown for $C_L = 1.0$ in Figures 8.5 and 8.6.

The use of vectored thrust in a turn

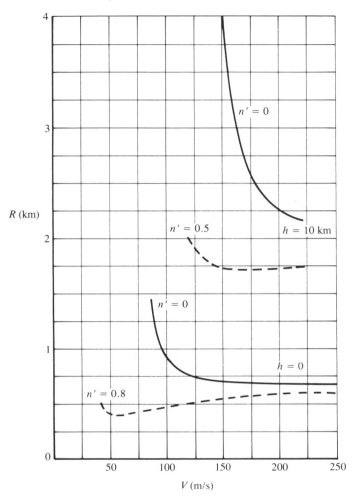

Figure 9.8. Reduction of minimum radius of turn by use of vectored thrust, with $w = 4 \text{ kN/m}^2$ and C_L limited to 1.0.

The broken lines in Figure 9.8 show the radius for the same lift coefficient but with vectored thrust to give n' equal to 0.8 at sea level and 0.5 at a height of 10 km. At both these heights the use of vectored thrust gives a substantial reduction of turn radius, but the reduction becomes smaller as the speed is increased because this allows a larger value of n, so making n' a smaller proportion of $(n + n')$. The rate of turn can of course be found from these results by using Equation (8.15).

If the turn is to be maintained without loss of speed or height it is necessary that the component of thrust acting along the flight path should not be less than the drag. When this condition is just satisfied

$$F \cos \theta_F = D$$

or

$$f \cos \theta_F = D/W = n\beta. \qquad (9.12)$$

Equation (8.26) then shows that

$$f \cos \theta_F = n'/\tan \theta_F = \tfrac{1}{2}\beta_m(v^2 + n^2 v^{-2}). \qquad (9.13)$$

As an example, Figure 9.9 shows values of θ_F and f calculated from Equation (9.13) for the turns represented in Figure 9.8, assuming that the drag of the aircraft is given by Equation (3.6) with $K_1 = 0.025$ and $K_2 = 0.075$. As in Example 9.1 these values give $\beta_m = 0.0866$ and $C_L^* = (\tfrac{1}{3})^{1/2} = 0.5773$. The results given in Figure 9.9 show that the large reductions of turn radius shown for the lower speeds in Figure 9.8 can be achieved without loss of speed or height with values of θ_F similar to those shown for PJB flight in Figure 9.5, i.e. large enough to give $\sin \theta_F \approx 1$, so that the required values of f are not much greater than the values of n' ($= f \sin \theta_F$) specified in Figure 9.8. For the higher speeds the drag is increased and smaller values of θ_F are needed to

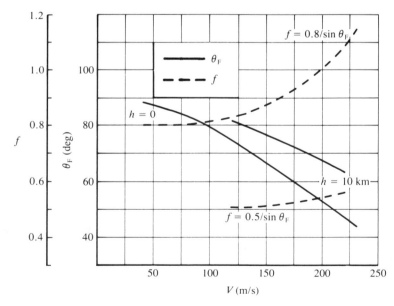

Figure 9.9. Values of θ_F and f required for turns shown in Figure 9.8, with no loss of speed or height ($K_1 = 0.025$, $K_2 = 0.075$).

make $F\cos\theta_F$ large enough to overcome the drag, so that for given values of n' the required thrust/weight ratios f are increased, the low-altitude example showing that the thrust requirement can even exceed that for hovering flight.

9.7 Other uses of vectored thrust in combat

If the angle θ_F as defined in Figure 9.3 can be increased to values well above 90° it is possible to generate a rearward component of thrust and this enables the aircraft to achieve even more of the hovering manoeuvrability normally associated with a helicopter. Of greater value than this is the potential for what has been called VIFF (Vectoring In Forward Flight). In this the pilot rotates the thrust nozzles to change θ_F while the forward speed is quite substantial and while the thrust component $F\sin\theta_F$ is not needed to support the weight of the aircraft. The consequent excess of total effective lift and sudden loss of opposition to drag produce a rapid and unexpected departure from the previous flight path and previous speed. Such a manoeuvre can be valuable in combat, whether it is used in association with a turn or in nearly straight flight.

10

Transonic and supersonic flight

On an aircraft in supersonic flight *shock waves* are always present, extending to a great distance from the aircraft. As mentioned in § 3.1.2 they are the cause of an additional component of drag, the *wave drag*, which has been neglected in the preceding chapters but which must be considered in transonic and supersonic flight as an important component of the total drag. The coefficient of wave drag of an aircraft depends on the lift coefficient and on the Mach number M and hence, with this component of drag included, the drag polar relating C_D to C_L now depends on M. The drag polar may still be represented approximately by the simple parabolic drag law as given by Equation (3.6) but now the coefficients K_1 and K_2 are functions of Mach number.

In the preceding chapters where wave drag was neglected, a curve relating β to V_e, as in Figure 2.9, could be applied to all heights, for a given aircraft weight, but with wave drag included this is no longer valid, because for any given value of V_e the Mach number varies with height. Thus no simplification can be obtained by the use of V_e, rather than V or M, and consequently the use of V_e^* and the speed ratio v is no longer helpful in the calculation of performance. In contrast, the quantity $(f - \beta)$ is still of prime importance in determining rate of climb and acceleration, as it is at lower speeds, and in general this quantity must be calculated for each combination of speed and height. As noted in § 5.2.4 the maximum thrust of a turbojet or turbofan engine varies substantially with speed in the transonic and supersonic range and this has important effects on the way in which $(f - \beta)$ varies with speed for a given height.

The transonic range of speed is usually taken to be that in which the air flow near to the aircraft is of mixed type, i.e. there are substantial regions where it is subsonic and others where it is supersonic. For most purposes this range of speed may be taken to include Mach numbers between about 0.8 and 1.2, although small regions of supersonic flow

Drag 253

can be formed at Mach numbers much lower than 0.8 when the lift coefficient is high. In considering transonic speeds in this chapter the emphasis will be on the lower end of the transonic speed range as defined here, between Mach numbers of about 0.8 and 0.95, because this is the range of Mach number in which, for most aircraft, the values of C_D and β begin the rapid rise which becomes quite steep as M approaches 1.0.

The additional drag that occurs in this transonic speed range is wave drag, caused by the formation of shock waves. As in supersonic flight, the coefficient of wave drag depends on both C_L and M, so that the drag polar depends on M, with consequences for performance that are to be outlined below. In particular, the conditions for economical cruising will be re-examined and it will be shown that for estimations of acceleration and climb performance, rate of turn and range in supersonic flight, a remarkable amount of the simple analysis normally used for subsonic speeds can be extended for use at transonic and supersonic speeds.

10.1 Drag

As already noted, flight at transonic or supersonic speed generates an additional component of drag, the wave drag, and in accordance with Figure 3.1 it is convenient to divide this into two components, one independent of lift and the other dependent on lift. Küchemann (1978) has considered in some detail the principles underlying the generation of wave drag at supersonic speeds and has shown that at any given Mach number the lift-dependent wave drag coefficient is approximately proportional to C_L^2. The vortex drag is generated in the same way as at subsonic speeds, so the coefficient C_{DV} is also closely proportional to C_L^2, and the relatively small remaining component of lift-dependent drag coefficient, due to viscous effects, may be assumed again to be proportional to C_L^2, as it was in Chapter 3. Thus the total lift-dependent drag coefficient is approximately proportional to C_L^2 and the overall drag coefficient may therefore be represented by Equation (3.6), but it is important to note that each of the coefficients K_1 and K_2 in this equation is now a function of Mach number. Küchemann (1978) has given a comprehensive account of the aerodynamic design methods that are available for achieving low values of C_D and β at transonic and supersonic speeds.

Although at supersonic speeds almost all of the wave drag is directly associated with the loss of available energy through generation of shock waves, the situation is often different in the transonic regime at Mach numbers in the region of 0.8 or 0.9 because there may then be an important additional component of wave drag caused by separation of the boundary layer induced by a shock wave. At Mach numbers up to about 0.9 this *shock-induced separation drag* is often the greater part of the total wave drag. (It can be argued that this separation drag should be included in the viscous drag of Figure 3.1, but it is considered here as part of the wave drag because it cannot exist unless shock waves are present.) At high subsonic flight speeds, where shock-induced separation drag is relatively large, it is hardly to be expected that this component of the drag coefficient will be proportional to C_L^2 and it is often found that Equation (3.6) becomes progressively less satisfactory as the Mach number increases above about 0.8, even when the most suitable values of K_1 and K_2 are chosen for each value of M.

It is sometimes assumed that in subsonic flight, at a Mach number low enough to ensure the absence of shock waves, the drag polar relating C_D to C_L will be independent of M. Although this is usually a good approximation there is no exact theoretical basis for it, because variation of Mach number changes the pressure distribution on the aircraft, even when there are no regions of supersonic flow and no shock waves, and this can change the viscous drag. Measurements in wind tunnels on aerofoils and aircraft models often show a very gradual increase of C_D with M at constant C_L, in a range of Mach number from about 0.6 or 0.7 up to the value at which C_D starts to rise steeply.

For a subsonic transport aircraft powered by turbofans it has been shown in §7.5 that maximum range is obtained when the cruising Mach number is close to the drag divergence Mach number M_D at which the drag coefficient starts to rise steeply. Hence in considering variations of C_D with Mach number for this class of aircraft the range of M that is of most interest lies near to and just above M_D. As a typical example of the dependence of C_D on C_L and M, Figure 10.1 shows drag polars for the Boeing 747 as given by Hanke & Nordwall (1970). In order to explore the validity of Equation (3.6) the drag coefficients for three Mach numbers have been replotted against C_L^2 in Figure 10.2, with straight lines giving the best fit in the range of C_L from 0.3 to 0.5. (The straight lines drawn in the figure show that this is the important range of C_L for cruising, because they give values of C_L^* of 0.52 for $M = 0.7$ and 0.47 for $M = 0.86$, and with $v > 1$ the value of C_L in the cruise is always lower than C_L^*.) The straight lines fit the data given by Hanke & Nordwall (1970) very well in the specified range of C_L at $M = 0.7$ and reasonably well at $M = 0.86$, but at $M = 0.92$ there is no part of the $C_D - C_L^2$ curve that is straight and Equation (3.6) fails to represent the $C_D - C_L$ relation adequately. As explained earlier, this failure may be attributed to the large component of drag caused by shock-induced separation, but the Mach number of 0.92 is well above the optimum value for economical

Drag

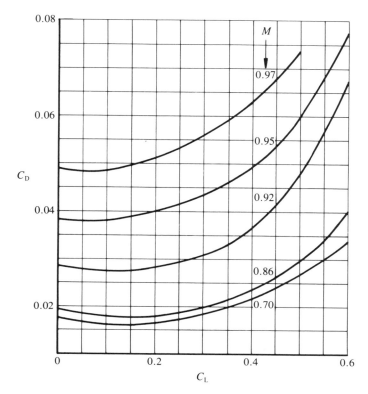

Figure 10.1. Drag polars for Boeing 747, from Hanke & Nordwall (1970).

cruising and the failure of Equation (3.6) at this high Mach number is therefore of no importance in the calculation of cruising performance for a subsonic transport.

Figure 10.3 is derived from Figure 10.1 and shows C_D plotted against M for several fixed values of C_L. This shows that for the Boeing 747 the Mach number M_D at which C_D starts to rise rapidly is well above 0.8 for all lift coefficients up to 0.5, but there is some tendency for M_D to decrease as C_L rises. (On some other aircraft the decrease of M_D with increasing C_L is more pronounced.)

As shown in earlier chapters the drag/lift ratio β is more relevant to performance calculations than the drag coefficient C_D by itself and it is therefore of interest to consider how β varies as the Mach number increases. At constant C_L the curves relating β to M would be essentially the same as those shown in Figure 10.3, but as the Mach number increases in level flight the lift coefficient does not remain constant. The required relation between M and C_L is found by

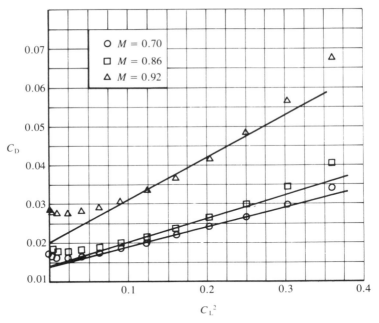

Figure 10.2. $C_D - C_L^2$ relation for Boeing 747, from Hanke & Nordwall (1970). Straight lines give best fit for $0.3 \leq C_L \leq 0.5$.

considering the equation for the wing loading

$$w = \tfrac{1}{2}\rho V^2 C_L,$$

from which Equation (1.15) shows that

$$w = \tfrac{1}{2}\gamma_a p M^2 C_L = 0.7 p_0 \delta M^2 C_L,$$

giving

$$M^2 C_L = (w/\delta)/(0.7 p_0) = 1.41 \times 10^{-5} \times w/\delta, \qquad (10.1)$$

with w in N/m². Thus $M^2 C_L$ remains constant as M increases and any specified value of $M^2 C_L$ corresponds to a value of w/δ given by Equation (10.1). For example, if $M^2 C_L = 0.3$ it follows that $w/\delta = 2.13 \times 10^4$ N/m² and if $w = 5000$ N/m² the pressure ratio δ must be 0.235, corresponding to a height of 10.7 km.

In Figure 10.4 the full lines are based on the data shown in Figures 10.1 and 10.3 and show the relation between β and M for several constant values of $M^2 C_L$. The broken line is the envelope of these curves and shows β_m, the minimum value of β, as a function of M. For low Mach numbers β_m is seen here to be independent of M as assumed in earlier chapters.

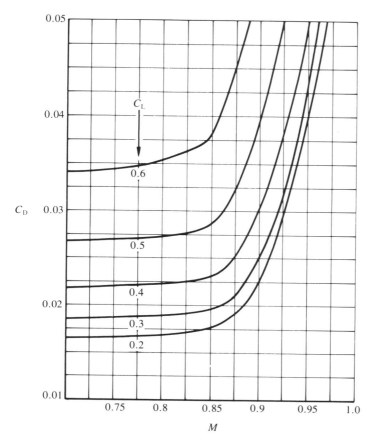

Figure 10.3. Drag of Boeing 747 at transonic speeds, from Hanke & Nordwall (1970).

The variation of C_D with M and C_L for a supersonic aircraft will be illustrated by considering two very different aircraft as examples. The first of these is Concorde (Figure 10.5), a supersonic civil aircraft designed for the best possible fuel economy in the cruise, together with acceptable take-off and landing performance and reasonable fuel economy in the climb and acceleration to the cruising speed and height. Concorde has a slender wing with *controlled* flow separation at the leading edge, an important design concept which has been fully explained by Küchemann (1978).

The aircraft of the second example is the English Electric Lightning, an early design of supersonic combat aircraft with a conventional wing having 60° sweepback at the leading edge. For this aircraft, as for other combat designs, the requirements include high rates of climb and acceleration, together with high maximum speed and good turning performance at all heights. In some cases the operational requirements for aircraft of this class imply that fuel economy is more important at high subsonic than at supersonic speeds; there will then be less attention paid to aerodynamic refinement for low drag at supersonic speeds than in the case of Concorde.

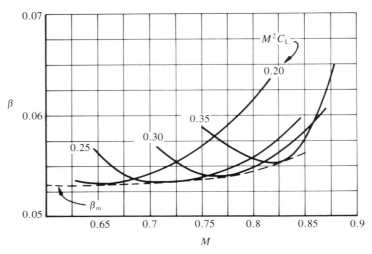

Figure 10.4. Drag/lift ratio β for Boeing 747, from Hanke & Nordwall (1970).

Figure 10.5. The Concorde supersonic civil transport cruises at $M = 2.0$ and at heights from 15 to 18 km. The slender wing is designed to give *controlled* flow separation in all conditions of flight.

Drag

Figure 10.6 shows an approximate representation of the drag coefficient of Concorde, derived from data supplied by British Aerospace plc for a range of C_L at several Mach numbers, and using linear interpolation between these Mach numbers. With these linear interpolations the values of C_D are not exact but they are sufficiently accurate for use in approximate performance calculations. In particular, the figure shows the steep rise of C_D at Mach numbers close to 1, which is caused by the formation of shock waves and is characteristic of all aircraft.

At any given Mach number, the Concorde drag data show a nearly linear variation of C_D with C_L^2, in accordance with Equation (3.6), and this is shown for three values of M in Figure 10.7. It should be noted that for this aircraft, unlike the Boeing 747 considered earlier, the linear relation is maintained through the transonic speed range because there is controlled flow separation at the leading edge at all Mach numbers, with no change in the general nature of the flow as M increases.

From straight lines of the kind shown in Figure 10.7, values of the coefficients K_1 and K_2 in Equation (3.6) have been derived and are shown in Figure 10.8, using linear interpolation between Mach numbers as explained earlier. As the Mach number rises from 1.2 to 2.0 there is a substantial increase of K_2 caused by an increase of lift-dependent wave drag, but in the same range of M there is a slight decrease of K_1.

The Equation (10.1) given earlier for level flight is valid also for any

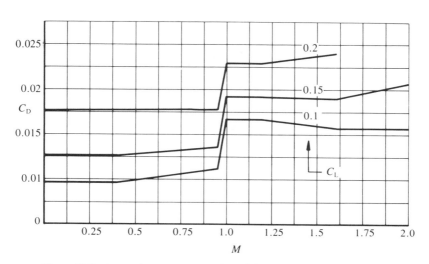

Figure 10.6. Approximate representation of drag coefficient of Concorde.

small angle of climb for which $\cos \gamma \approx 1$. This equation shows that for any given wing loading and height, $M^2 C_L$ remains constant and the full lines in Figure 10.9 show how β varies with Mach number for three constant values of $M^2 C_L$. Equation (10.1) shows that if the wing loading w is expressed in kN/m^2 the ratio $w/\delta = 70.9 M^2 C_L$ and hence the values of w/δ for the three curves shown as full lines in Figure 10.9 are 17.7, 28.4 and 42.6 kN/m^2. If the assumption that $\cos \gamma = 1$ is not valid the equation to be used in place of Equation (10.1) is

$$M^2 C_L = 1.41 \times 10^{-5} \times (w/\delta) \cos \gamma \tag{10.2}$$
(with w in N/m^2)

and for any given value of $M^2 C_L$ the ratio w/δ is then increased by the factor $\sec \gamma$.

From the values of K_1 and K_2 given in Figure 10.8, values of C_L^* and β_m have been calculated from Equations (3.12) and (3.13) and these are shown by the broken lines in Figure 10.9. The increase of K_2 with M in the supersonic speed range leads to a decrease of C_L^* and some increase of β_m, even though K_1 decreases slightly. It should be emphasised that although β has the minimum value β_m when $C_L = C_L^*$, for any given Mach number, the variation of β with speed is

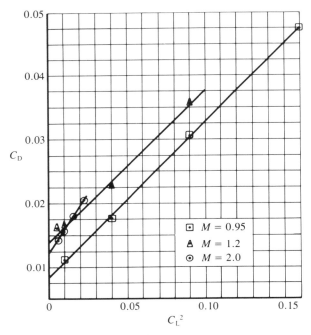

Figure 10.7. $C_D - C_L^2$ relation for Concorde.

Drag

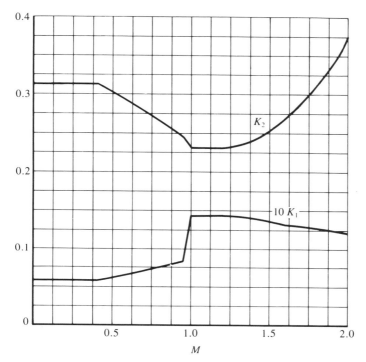

Figure 10.8. Approximate representation of coefficients K_1 and K_2 for Concorde.

not given by Equation (3.18), because in general K_1, K_2 and β_m all vary with Mach number. The curve in Figure 10.9 relating β_m to M is the envelope of the curves relating β to M at constant $M^2 C_L$, as it is in Figure 10.4, and for any specified Mach number there is a value of $M^2 C_L$ which gives $C_L = C_L^*$ and $\beta = \beta_m$.

As a further example, Figure 10.10 shows the drag coefficient of the Lightning for a wide range of lift coefficients. Comparison with Figure 10.6 shows that at supersonic speeds C_D is much greater for the Lightning than for Concorde, reflecting the much earlier design of the Lightning and the very different requirements for the two aircraft.

At Mach numbers below 1.2, the relation between C_D and C_L^2 for the Lightning is not a simple linear one and this is illustrated in Figure 10.11, which shows the relation for $M = 0.3$. There is discontinuity in slope at the point B, which is caused by a change in the nature of the flow separation on the sweptback wing, a change which does not occur on Concorde, with its slender wing and controlled flow separation. As shown in Figure 10.12, the lift coefficient C_{LB} at the 'break point' B falls progressively as M increases up to 1.2 and is zero for $M \geq 1.2$, i.e.

the segment BC in Figure 10.11 increases in length while the segment AB is reduced and the curve becomes a single straight line with no break for $M \geq 1.2$.

Values of K_1 and K_2 can be derived from the straight lines relating C_D to C_L^2 and these are shown in Figure 10.13, where the full and broken lines refer respectively to $C_L > C_{LB}$ and $C_L < C_{LB}$. The coefficient K_2 increases rapidly as M rises above about 1.2, because of the large increase of lift-dependent wave drag. The most notable feature of the K_1 curve, as for Concorde, is the steep rise at Mach numbers close to 1, followed by a slight decrease at higher Mach numbers.

Figure 10.14 is constructed in the same way as Figure 10.9 for Concorde; the full lines give β for the three constant values of $M^2 C_L$ that are shown, and the broken lines give C_L^* and β_m. The values of w/δ for the three values of $M^2 C_L$ are of course the same as those given earlier for Concorde, provided it can be assumed that $\cos \gamma \approx 1$. For the Lightning, the increase of β_m and reduction of C_L^* as M increases up to 2.0 are more pronounced than for Concorde, because of the greater rate of increase of lift-dependent wave drag and hence

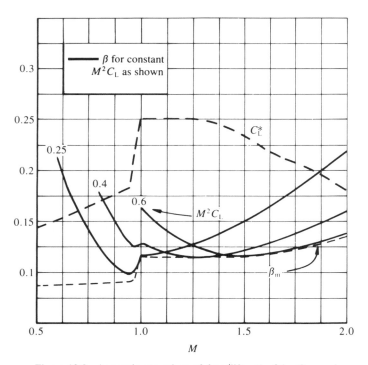

Figure 10.9. Approximate values of drag/lift ratio β for Concorde.

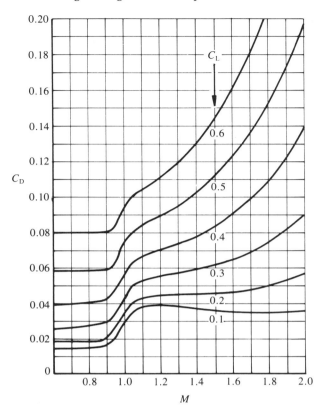

Figure 10.10. Drag coefficient of Lightning.

of K_2. As in Figures 10.4 and 10.9 the broken line showing β_m is the envelope of the curves giving β at constant $M^2 C_L$.

10.2 Range at high subsonic speeds

In the discussion of range in Chapter 7 it has been assumed that for any given C_L the drag coefficient C_D is independent of Mach number, provided that this is less than the value M_D at which C_D starts to rise steeply. With this assumption it was found that the specific range would be maximum with $\beta = \beta_m$ and $M = M_D$, but the curves in Figure 10.3 show that it is difficult to define M_D precisely and the slope dC_D/dM increases progressively as M rises from about 0.7. The characteristics shown for the Boeing 747 in Figure 10.3 are typical of long-range subsonic civil aircraft and there is therefore a need to examine in more detail the flight conditions required for maximum range.

264   *Transonic and supersonic flight*

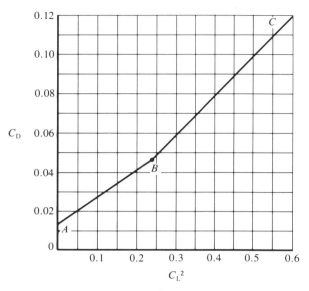

Figure 10.11. C_D–C_L^2 relation for Lighting at $M = 0.3$.

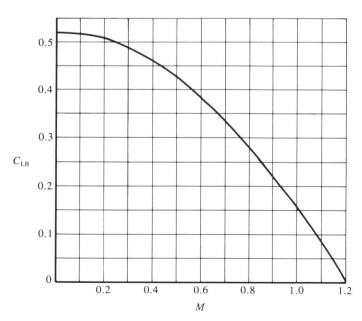

Figure 10.12. Lift coefficient at 'break' in C_D–C_L^2 relation for Lightning.

Range at high subsonic speeds

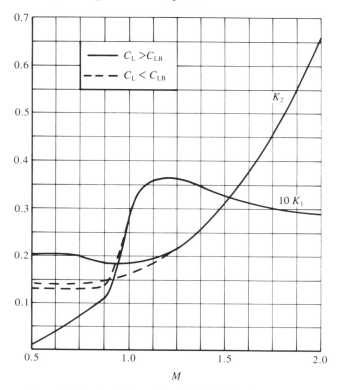

Figure 10.13. Coefficients K_1 and K_2 for Lightning.

For maximum cruising range, the specific range r_a as defined in Equation (7.1) should be maximum at every instant of the cruise and this requires minimum $c\beta/V$. In order to simplify the argument it will be assumed that the air temperature is constant, so that $V \propto M$. Thus $c\beta/M$ is required to be minimum and if c is assumed to be independent of speed and height the requirement is for minimum β/M.

Taking the Boeing 747 as an example and using the data given in Figure 10.4, Figure 10.15 shows β/M plotted against M for several constant values of $M^2 C_L$, representing constant values of w/δ as given by Equation (10.1). For a given value of the wing loading w, each of the curves represents a constant value of δ and therefore a constant height. For each of the curves, the position of the minimum gives the Mach number for maximum r_a at this height and comparison with Figure 10.4 shows that this Mach number for maximum r_a is greater than the value for minimum β, especially for the lower values of $M^2 C_L$ where the minimum β occurs at a relatively low Mach number.

If there is a free choice of height and speed, and no limitations are imposed by the available thrust, maximum r_a is obtained by operating at the lowest of the minima shown in Figure 10.15, i.e. with M^2C_L in the region of 0.35 and $M \approx 0.83$. Equation (10.1) shows that this value of M^2C_L corresponds to $w/\delta = 24.8 \text{ kN/m}^2$ and this implies a height of about 11.7 km if $w = 5 \text{ kN/m}^2$. If a cruise-climb is allowed, this same value of w/δ can be maintained throughout the cruise.

It was shown in § 7.5.4 that reduction of height allowed the aircraft to cruise with a lower value of F/σ, and thus with a lower engine weight, with only a small penalty in specific range. Figure 10.15 can be used to show that there is a similar effect in the case considered here, even if there is no change of Mach number. Still assuming that the air

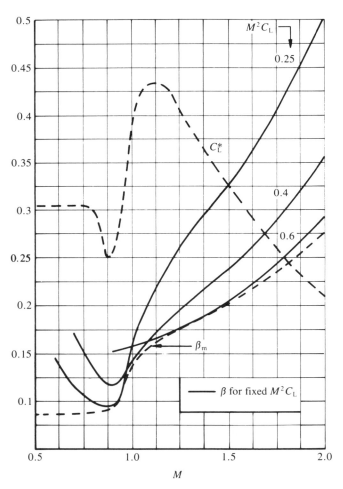

Figure 10.14. Drag/lift ratio β for Lightning.

Range at high subsonic speeds

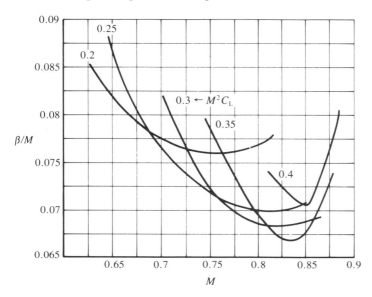

Figure 10.15. Ratio β/M for Boeing 747, from data given by Hanke & Nordwall (1970).

temperature is constant, so that $V \propto M$,

$$L = \tfrac{1}{2}\rho_0 \sigma V^2 S C_L,$$

the required thrust is

$$F = D = \beta L$$

and

$$F/\sigma \propto \beta M^2 C_L.$$

As an example, if M is maintained at the constant value of 0.83, $M^2 C_L$ can be reduced from 0.35 to 0.3 by reducing height to give greater σ and smaller C_L. Figure 10.15 shows that β is then increased by only about 2.4%, yet F/σ is reduced in the ratio $(0.3/0.35) \times 1.024 = 0.88$. Still assuming constant c, the specific range is inversely proportional to β and is reduced by 2.4%. This is a rather greater range penalty than the 1% shown in Figure 7.2 for the same 12% reduction of F/σ at a lower Mach number.

Instead of assuming that c is independent of speed and height, a more realistic assumption is that c is given by Equation (5.11), and it is shown in Figures 5.10 and 5.11 that for a typical modern civil turbofan the exponent n in that equation is about 0.5. Assuming that $c \propto M^{1/2}$, the condition for maximum r_a is that $\beta/M^{1/2}$ should be minimum and

Figure 10.16 shows this quantity plotted in the same way as β/M was given in Figure 10.15. The conclusions from Figure 10.16 are qualitatively the same as those given earlier for constant c, but for the three lowest values of $M^2 C_L$ the minimum now occurs at a lower Mach number than that shown in Figure 10.15. With a free choice of speed and height and with the more realistic assumption about fuel consumption the conditions for maximum r_a are almost the same as those given by Figure 10.15 for constant c, viz $M \approx 0.83$ and $M^2 C_L$ in the region of 0.35.

10.3 Climb and acceleration in supersonic flight

The analysis of climbing performance in Chapter 4 depended on the simple parabolic drag law as expressed by Equation (3.6), with the coefficients K_1 and K_2 having constant values for all speeds and heights. In supersonic flight, and also at high subsonic speeds, both K_1 and K_2 vary with Mach number and the analysis given earlier cannot be used. Also, the discussion given in § 4.1 of the approximation $\cos \gamma = 1$ is not valid if K_1 and K_2 are variable, and unless γ is obviously small it is necessary to insert the correct value of $\cos \gamma$ in the equations, using iteration if necessary. It will be shown in Example 10.2, however, that even with a high-performance combat aircraft, having very high values of both the thrust and the *rate* of climb, the *angle* of climb is usually fairly small at supersonic flight speeds.

When an aircraft climbs at a supersonic speed it often accelerates at the same time and it is therefore convenient to express the performance in terms of the specific excess power, using Equation (4.32) to relate this to acceleration and rate of climb. The specific excess power was defined in Chapter 4 as

$$P_{se} = V(F - D)/W = V(f - \beta \cos \gamma) \tag{10.3}$$

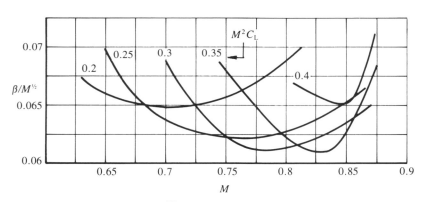

Figure 10.16. Ratio $\beta/M^{1/2}$ for Boeing 747, from data given by Hanke & Nordwall (1970).

and Equation (4.32) shows that

$$P_{se} = dh/dt + (V/g)\,dV/dt. \tag{10.4}$$

There are two important special cases:
(a) excess power allows a climb with zero acceleration (constant V): the climb rate becomes

$$V_c = dh/dt = P_{se} = V(f - \beta \cos\gamma), \tag{10.5}$$

(b) the excess power is directed entirely towards acceleration, with $\gamma = 0$ and no gain of height, giving an acceleration

$$dV/dt = gP_{se}/V = (f - \beta)g. \tag{10.6}$$

In cases where γ is not small enough to justify the assumption that $\cos\gamma = 1$, the $\cos\gamma$ term must be included in the Equations (10.3) and (10.5) and the lift coefficient must be obtained from Equation (10.2) and not (10.1). (The change of C_L caused by the $\cos\gamma$ term affects the value of β.)

The functional relation (5.5) is valid at supersonic as well as at subsonic speeds and shows that for a turbojet or turbofan engine F/δ depends only on M, if $N/\theta^{1/2}$ is constant. For $h > 11$ km in the ISA, the temperature is constant and hence for a fixed engine speed N a single curve relating F/δ to M determines the thrust at any speed and height. (For heights below 11 km a series of curves is needed, giving F or F/δ as a function of M for each height.) If, in addition, β is known as a function of C_L and M, either directly or through curves relating K_1 and K_2 to M, the climb and acceleration performance can be calculated from Equations (10.3) and (10.4).

As an example, Figure 10.17 shows the thrust and drag ratios f and

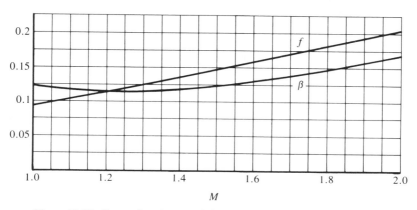

Figure 10.17. Concorde – thrust and drag ratios for $h = 12$ km, $m = 180\,000$ kg, no reheat.

β for Concorde at a height of 12 km. The mass is taken to be 180 000 kg, corresponding to take-off at the maximum mass of 185 000 kg with 5% of the fuel having been used in an accelerating climb. The f curve refers to the four Olympus 593 engines without reheat and is derived from Figure 5.17. The drag ratio β is obtained from the approximate values of K_1 and K_2 given in Figure 10.8, calculating C_L for the wing area of 358 m^2 by putting $\cos \gamma = 1$ in Equation (10.2). The greatest value of $(f - \beta)$ is only 0.037 and even with zero acceleration the corresponding angle of climb is only about 2°.

Figure 10.17 shows that for these values of the height and aircraft mass, $(f - \beta)$ and P_{se} are zero at a Mach number just above 1.2. In principle there should be another Mach number, higher than this, at which $P_{se} = 0$, and this would represent the maximum speed in steady level flight, as determined by the thrust and drag characteristics. There is, however, another factor which limits the speed of Concorde and many other supersonic aircraft, and this is the maximum allowable temperature of the aircraft structure (primarily aluminium alloy). The difference between this temperature and the ambient air temperature is approximately proportional to M^2 and the effect of this on Concorde is to limit the cruising Mach number to a value between 2.0 and 2.05, depending on the air temperature. In fact the drag data presented in Figure 10.17 are based on Figure 10.6 and represent only an approximation to the true drag. Whereas the upper convergence of the two curves would appear to be at a high Mach number, the two do in fact converge again at a relatively modest value of M.

A notable feature of Figure 10.17 is the increase of $(f - \beta)$ with flight speed. Reference to § 2.7 and Figure 2.12 shows that this is an indication of potential speed instability, for although Figure 2.12 shows negative slopes of the f and β curves in the region of instability ($V_e < V_{ec}$), whereas both these slopes are positive for the higher Mach numbers in Figure 10.17, the essential feature indicating possible speed instability is an increase of $(f - \beta)$ with speed. The speed instability thus predicted for Concorde in supersonic flight is also observed on other supersonic aircraft and is caused by the rapid increase of thrust with Mach number, more rapid in this case than the increase of β. The instability does not cause any difficulty in practice, because Mach number and pitch attitude are among the flight parameters controlled by an autopilot throughout the supersonic flight of Concorde and other aircraft.

Figure 10.17 shows that the Olympus 593 engines in Concorde do not generate sufficient thrust without reheat to reach a height of 12 km at Mach numbers between 1.0 and 1.2. The reheat system in these engines increases the thrust by about 25% and is normally used in the

Climb and acceleration in supersonic flight

accelerating climb from ($M = 0.95$, $h = 8$ km) to ($M = 1.7$, $h = 13$ km). In any case, whether reheat is used or not, the thrust increases substantially as height is reduced and, for the heights used, reheat ensures that ample thrust is available for the accelerating climb. Reheat is used for a relatively large part of the accelerating climb in order to complete this part of the flight path as quickly as possible, thus minimising the use of fuel even though the sfc is increased by 25–30% when reheat is used.

EXAMPLE 10.1. SUPERSONIC CIVIL AIRCRAFT: MAXIMUM VALUES OF RATE OF CLIMB AND ACCELERATION

A Concorde aircraft with a mass of 180 000 kg is flying at $M = 1.8$ and $h = 12$ km. At this Mach number the coefficients in Equation (3.6) are $K_1 = 0.0128$, $K_2 = 0.310$ as shown in Figure 10.8, and for each of the four engines $F/\delta = 422$ kN (without reheat). The wing area is 358 m². Find the angle of climb and the rate of climb with zero acceleration; find also the acceleration in level flight.

Some preliminary calculations are needed to find f and β.

$$W = 180 \times 9.81 = 1766 \text{ kN}$$

and

$$w = W/S = 1766/358 = 4.93 \text{ kN/m}^2.$$

For $h = 12$ km the pressure ratio is $\delta = 0.1908$ and the thrust ratio then becomes $f = F/W = 4 \times 422 \, \delta/W = 0.1824$.

Assuming that $\cos \gamma = 1$ and using Equation (10.2),

$$C_L = 0.0141 \times 4.93/(M^2\delta) = 0.1124.$$

Hence

$$\beta = K_1/C_L + K_2 C_L = 0.1487$$

and

$$P_{se} = V(f - \beta) = 0.0337 \, V.$$

For $h > 11$ km, Appendix 2 gives

$$a = a_0(a/a_0) = 340.3 \times 0.8671 = 295.1 \text{ m/s}$$

and for $M = 1.8$, $V = 1.8a = 531.2$ m/s.

With zero acceleration, Equation (10.5) gives the rate of climb as

$$V_c = V \sin \gamma = P_{se} = 17.9 \text{ m/s}.$$

Hence

$$\sin \gamma = P_{se}/V = 0.0337 \quad \text{and} \quad \gamma = 1.93°.$$

Since this is the greatest possible angle of climb, unless there is a negative acceleration, the assumption that $\cos \gamma = 1$ is fully justified.

For level flight, Equation (10.6) gives

$$\text{acceleration} = dV/dt = gP_{se}/V = 0.0337g = 0.331 \text{ m/s}^2.$$

(It may be noted that this acceleration would be imperceptible to a passenger, and indeed the aircraft attitude required to obtain the climb angle of 1.93° would probably also be imperceptible.)

EXAMPLE 10.2. SUPERSONIC COMBAT AIRCRAFT: MAXIMUM VALUES OF RATE OF CLIMB AND ACCELERATION

A combat aircraft with a wing loading of 4 kN/m² is flying at $M = 1.8$ and $h = 12$ km. The coefficients in Equation (3.6) are $K_1 = 0.025$, $K_2 = 0.4$ and the maximum thrust is $0.5W$. Find the angle of climb and the rate of climb with zero acceleration; find also the acceleration in level flight.

As in Example 10.1, $V = 531.2$ m/s and $\delta = 0.1908$.

For level flight, Equation (10.1) gives

$$C_L = 0.0141 \times 4/(M^2 \delta) = 0.0912$$

and hence

$$\beta = K_1/C_L + K_2 C_L = 0.3106.$$

Then, since $f = 0.5$, $P_{se} = V(f - \beta) = 0.1894 V$ and Equation (10.6) gives the acceleration in level flight as

$$dV/dt = gP_{se}/V = 0.1894g = 1.858 \text{ m/s}^2.$$

For the climb with zero acceleration it will be assumed, as a first approximation, that $\cos \gamma = 1$. Then P_{se} has the value already found and the rate of climb is

$$V_c = V \sin \gamma = P_{se} = 0.1894V,$$

giving

$$\sin \gamma = 0.1894, \qquad \gamma = 10.92° \quad \text{and} \quad \cos \gamma = 0.982.$$

Then, using this value of $\cos \gamma$ in Equation (10.2), a more accurate estimate of the lift coefficient is

$$C_L = 0.0912 \cos \gamma = 0.0895,$$

giving

$$\beta = 0.3150 \quad \text{and} \quad P_{se} = V(f - \beta) = 0.1850 \, V.$$

Hence

$$\sin \gamma = P_{se}/V = 0.1850, \qquad \gamma = 10.66°$$

and $\cos \gamma = 0.983$.

This value of cos γ is so near to the value found earlier that no further iteration is needed and with zero acceleration the rate of climb is $0.1850V = 98.3$ m/s and the angle of climb is $10.66°$.

It should be noted that even for this aircraft, with a very high thrust/weight ratio for this altitude and a high rate of climb, the achievable angle of climb is only about $11°$ and the cos γ term in the equations gives only a small correction. With lower thrust the angle of climb would of course be smaller. A further point to be noted is that the few aircraft which have sufficient thrust at low altitudes to allow vertical flight shortly after take-off will not be able to sustain a climb angle of $90°$ up to high altitudes, because of the decrease of maximum thrust as height is gained.

10.4   Range at supersonic speeds

The specific range r_a has been defined in Equation (7.1) and integration of this over the cruise gives the total cruise range R_{ac} as in Equation (7.2) or (7.3). The range R_{ac} is maximum when the specific range r_a has the greatest possible value at every instant of the flight and the condition for this is that $c\beta/V$ should always be minimum. Since supersonic aircraft normally fly at very high altitudes when long range is required, the argument will be simplified by assuming that the height is always greater than 11 km, so that the air temperature is constant and $V \propto M$. Then for maximum specific range $c\beta/M$ is required to be minimum.

The functional relation (5.7) shows that the specific fuel consumption c is independent of height and depends only on M, if the engine speed and air temperature remain constant. Figure 5.13 shows that for the dry military turbofan 'A' there is a rise of c with Mach number, the rise becoming steeper as M increases, and for Mach numbers between about 1.6 and 2.0 the ratio c/M is nearly independent of M. For the engine with maximum reheat, Figure 5.14 shows a more complex variation of c with M, leading to a reduction of c/M as M rises throughout the speed range. For the (dry) Olympus 593 engine in Concorde, Figure 5.17 shows that increasing M causes only a small rise of c, so that c/M decreases.

For any given Mach number the ratio $c\beta/M$ is minimum when β has the minimum value for that Mach number. Since Equation (3.6) is usually valid at supersonic speeds, β has the minimum value $\beta_m = 2(K_1 K_2)^{1/2}$ when $C_L = C_L^* = (K_1/K_2)^{1/2}$, but in general K_1, K_2, β_m and C_L^* all vary with Mach number, as seen in Figures 10.8 and 10.13.

If K_1 and K_2 are known as functions of Mach number, C_L^* can be found for any specified Mach number and with $C_L = C_L^*$ Equation (10.1) can be used to find the value of w/δ giving minimum β (and hence minimum $c\beta/M$) in level flight. Thus for any given value of the wing loading w the height required for maximum r_a can be found. At this optimum height the specific range is

$$r_a = V/(c\beta_m W)$$

and hence

$$mr_a = Ma/(c\beta_m g), \qquad (10.7)$$

where a is the velocity of sound.

As an example, Figure 10.18 shows mr_a for Concorde at the optimum height, calculated from Equations (10.1) and (10.7) by using the approximate values of K_1 and K_2 given in Figure 10.8 and the values of c given in Figure 5.17 for the Olympus 593 engine. The figure shows that for this aircraft the specific range r_a increases strongly with Mach number in the range from 1.0 to about 1.8 and maximum range is obtained at the normal operating cruise Mach number of about 2. The rise of r_a with Mach number is also found on many other supersonic aircraft and for these aircraft, as for Concorde, the lower supersonic range of speed is uneconomic and cruise at the higher Mach numbers is preferable.

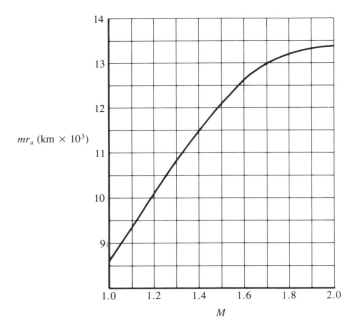

Figure 10.18. Concorde – specific range r_a for aircraft mass m. Height is chosen to give maximum r_a for each Mach number.

Range at supersonic speeds

It was shown in § 7.5.4 and in Figures 7.1 and 7.2 that for a subsonic aircraft an increase of v above 1, i.e. a reduction of C_L below C_L^*, could give a substantial reduction of F/σ with only a small reduction of r_a. In the subsonic case the reduction of C_L is best achieved by reducing the height, while keeping the Mach number close to the drag divergence value M_D. A similar effect can be obtained at supersonic speeds; reduction of C_L below C_L^* by reducing the height, at constant Mach number, gives a substantial reduction of F/σ (and F/δ) at the expense of only a small reduction of range. This is shown in Figure 10.19 for Concorde at $M = 2.0$, where mr_a has been calculated as for Figure 10.18, but with β_m in Equation (10.7) replaced by the correct value of β as given by Equation (3.19) for each value of C_L/C_L^*. The required value of F/δ has been found from the equation $F = \beta W$ and it should be noted that the large decrease of F/δ as C_L/C_L^* falls below 1 occurs because $\delta \propto C_L^{-1}$ at constant M and there is only a small increase of β as C_L falls. Figure 10.19 indicates that the available thrust from the Olympus 593 engines is insufficient to maintain level flight at $M = 2.0$ with a lift coefficient greater than $0.76 C_L^*$. At this value of C_L the quantity mr_a is reduced from its maximum value of

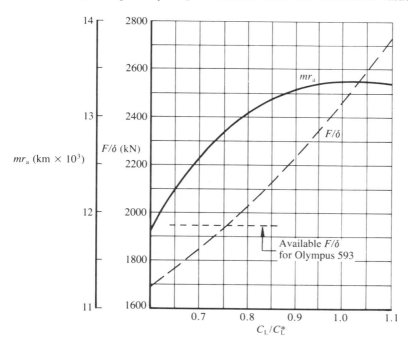

Figure 10.19. Approximate values of specific range and thrust for Concorde in level flight at $M = 2.0$.

276   *Transonic and supersonic flight*

13.37×10^3 km to 12.89×10^3 km, i.e. by 3.6%, but the result is inaccurate because it is based on drag data that are only approximate. The cruise condition actually used for Concorde corresponds to a ratio $C_L/C_L{}^*$ that is greater than 0.76 and gives only a very small reduction of r_a from the maximum value, while still allowing F/δ to be significantly less than the value required for $C_L = C_L{}^*$.

If the aircraft flies in a cruise-climb at a constant Mach number, so that $\sigma \propto \delta \propto m$, the values of C_L, β, F/δ and mr_a remain constant during the cruise and the total cruise range can be found by integration as shown in Equation (7.2). A more usual flight procedure is a stepped cruise consisting of two phases of level flight, with a climb from the lower to the upper height when approximately half the total cruise distance has been flown. As at subsonic speeds, the cruise range given by a stepped cruise is usually only slightly less than that obtained with a cruise-climb.

As mentioned in §1.1, there are two further factors affecting the cruise performance of a supersonic aircraft flying at high altitude. The first is the reduction of g due to the increased distance from the centre of the earth; this amounts to 0.6% at the typical Concorde cruise height of 18 km. The second is the upward centrifugal force due to the flight path following the earth's curved surface; this is greatest for an eastward flight over the equator, when it reduces the required lift by 1.8% at $M = 2$.

EXAMPLE 10.3. RANGE OF SUPERSONIC CIVIL AIRCRAFT

A supersonic civil aircraft with a wing area of 350 m² cruises at $M = 2.0$. At this Mach number the coefficients in Equation (3.6) are $K_1 = 0.012$, $K_2 = 0.35$ and the specific fuel consumption c is 3.3×10^{-5} kg/N s. In a cruise-climb at a constant lift coefficient, chosen to give maximum range, find the cruise range R_{ac} if the mass of the aircraft at the start of the cruise is 180 000 kg and 65 000 kg of fuel is available for the cruise. Also find the initial and final heights and the required value of F/δ.

By what percentages are R_{ac} and F/δ reduced if the lift coefficient is reduced to 85% of the value giving maximum range?

From Equations (3.12) and (3.13),

$$\beta_m = 2(K_1 K_2)^{1/2} = 0.1296$$

and

$$C_L{}^* = (K_1/K_2)^{1/2} = 0.1852.$$

As in Example 10.1, the velocity of sound is $a = 295.1$ m/s. For

Turning in supersonic flight 277

maximum range $C_L = C_L{}^*$, $\beta = \beta_m$ and Equation (10.7) gives

$$mr_a = (2 \times 295.1)/(0.033 \times 0.1296 \times 9.81) \text{ km} = 14\,067 \text{ km}.$$

Equations (7.2) and (7.3) then show that the cruise range is

$$R_{ac} = -\int_{m_1}^{m_2} r_a \, dm = -14\,067 \int_{m_1}^{m_2} (1/m) \, dm$$
$$= 14\,067 \ln(m_1/m_2)$$
$$= 6032 \text{ km}.$$

Equation (10.1) shows that

$$w/\delta = 70.9 M^2 C_L = 52.52 \text{ kN/m}^2$$

and if suffixes 1 and 2 are used to denote the start and end of the cruise,

$$w_1 = 5.045 \text{ kN/m}^2 \quad \text{and} \quad w_2 = 3.223 \text{ kN/m}^2.$$

Hence $\delta_1 = 0.0961$ and $\delta_2 = 0.0614$, for which the corresponding heights are $h_1 = 16.34$ km and $h_2 = 19.17$ km. F/δ remains constant and can be calculated from the conditions at the start of the cruise, giving

$$F/\delta = \beta_m m_1 g / \delta_1 = 0.1296 \times 180 \times 9.81/0.0961$$
$$= 2381 \text{ kN}.$$

If C_L is reduced to $0.85 C_L{}^*$, Equation (3.19) shows that $\beta = 1.013 \beta_m$ and hence the range is reduced by 1.3%. With no change of Mach number $\delta \propto C_L^{-1}$ and since $F = \beta W$ the new value of F/δ is

$$2381 \times 0.85 \times 1.013 = 0.861 \times 2381 \text{ kN}.$$

Thus F/δ is reduced by 13.9%.

10.5 Turning in supersonic flight

In a turn at a supersonic speed the Equation (8.7) is still valid but the limitations imposed by stalling or buffeting and by the available thrust need to be reconsidered, because of the influence of Mach number on the maximum usable lift coefficient and on the coefficients K_1 and K_2 in Equation (3.6).

The maximum usable lift coefficient for an aircraft at supersonic speeds is often determined by the onset of buffeting rather than conventional stalling. This upper limit of C_L is of the same order as that found at the higher subsonic speeds, usually lying in a range from about 0.7 to 1.0. As at subsonic speeds it varies with Mach number in a way that cannot be specified by any general equation or law. As an example, to illustrate the effects of C_L limitations on turning performance, Figure 10.20 refers to an aircraft flying at heights above 11 km. The full lines show the relations between the radius of turn R, the load

278   *Transonic and supersonic flight*

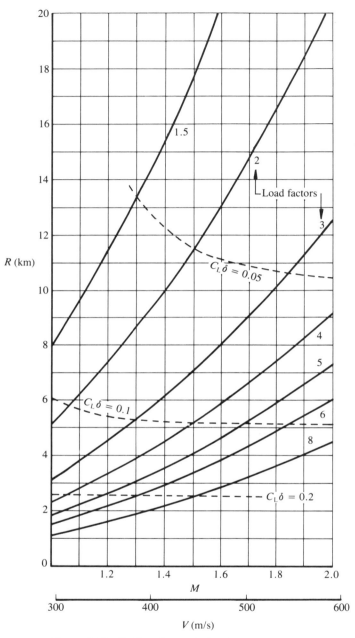

Figure 10.20. Radius of turn, limited by stall or buffet, $w = 4 \text{ kN/m}^2$, $h \geq 11$ km.

Turning in supersonic flight

factor n and the Mach number and have been derived from Equation (8.7), as in Figures 8.5 and 8.6. These curves are unaffected by variations of wing loading or height, provided that $h > 11$ km so that the velocity of sound is constant. The broken lines in Figure 10.20 represent constant values of $C_L \delta$ and have been calculated for a wing loading of 4 kN/m². Using C_{L0} to denote the lift coefficient in straight and level flight, as in Equation (8.9), the Equation (10.1) can be used to find $C_{L0}\delta$ in terms of w and M, then $C_L \delta$ is found from Equation (8.9) as $nC_{L0}\delta$. The lines of constant $C_L \delta$ in Figure 10.20 show that the minimum radius of turn becomes nearly independent of M and n when n is large. This feature also appeared in Figure 8.5 and has already been explained by consideration of Equation (8.14).

If the maximum usable C_L is known as a function of Mach number, the consequent limitations imposed on the radius of turn can be found from a diagram of the kind shown in Figure 10.20. For example, if the upper limit of C_L is 1.0 at $M = 1.84$, the parameter $C_L \delta$ is equal to 0.1 at $h = 16.1$ km (where $\delta = 0.1$) and is 0.05 at $h = 20.5$ km (where $\delta = 0.05$). Figure 10.20 then shows that the minimum radius of turn at this Mach number is 5.1 km at $h = 16.1$ km and is 10.6 km at $h = 20.5$ km, the maximum load factor being 6 at the lower height and 3 at the greater one. These numerical values also illustrate two further points:

(i) $n \propto \delta$ at constant M and C_L, because Equation (8.9) gives $n \propto C_{L0}^{-1}$ and Equation (10.1) gives $C_{L0}^{-1} \propto \delta$,
(ii) when n is large the minimum radius is nearly proportional to δ^{-1}, for the reason explained earlier in discussion of Equation (8.14).

As discussed in § 8.5 and 8.6 the condition required for an aircraft to turn without loss of height or speed is that P_{se} should not be less than 0, i.e. F should not be less than D. In the limiting condition, when height and speed can just be maintained, $P_{se} = 0$ and $F = D = n\beta W$, as shown by Equation (8.28), so that $f = F/W = n\beta$. For specified values of M, w, δ and n, the lift coefficient can be found as explained earlier from Equations (10.1) and (8.9). Then, if β_m and C_L^* are known as functions of Mach number, either directly or by calculation from K_1 and K_2, Equation (8.27) can be used to find β and hence $f (= n\beta)$ can be found for the limiting condition.

High wave drag at supersonic speeds often leads to large values of β_m, typically up to about two or three times the value found at subsonic speeds. These high values of β_m mean that even for moderate values of the load factor n the thrust required to maintain height and speed in a turn may be greater than the maximum thrust available from the engines, so that the turn can be made only with a loss of either height or speed, or both. As an indication of the thrust that may be required to maintain height and speed, Figure 10.21 shows values of f calculated for the Lightning at a height of 12 km, using the values of C_L^* and β_m given in Figure 10.14 and assuming a wing loading of 4 kN/m². The broken line shows the value of f required for straight and level flight, i.e. for $n = 1$, and this is of course the same as the value of β for $L = W$. In order to clarify the changes that occur in Figure 10.21 as n is increased, Figure 10.22 shows the variation with n of f, C_L/C_L^* and β/β_m for $M = 1.8$. The ratio C_L/C_L^* is of course directly proportional to n, because C_L^* and C_{L0} are constant and $C_L = nC_{L0}$. The ratio β/β_m is a function of C_L/C_L^*, as given by Equation (8.27), falling to the minimum value of 1 when $n \approx 2.75$ and $C_L/C_L^* = 1$. The thrust/weight ratio f is roughly proportional to n in the range of n for which β is nearly constant, but it increases less rapidly with n in the lower range of n for which β falls as n rises.

The Lightning is an early design of supersonic combat aircraft with a large wave drag. More modern designs have somewhat lower drag and would probably not require values of f as large as those shown in Figure 10.21, in order to maintain height and speed in a turn, but even with these modern aircraft the available thrust is usually an important factor limiting the turning performance.

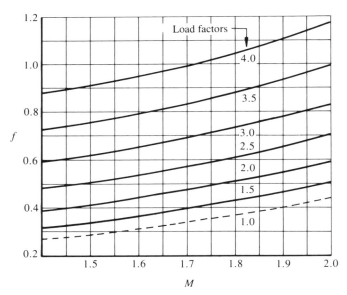

Figure 10.21. Thrust/weight ratio f required for a turn without loss of height or speed, $h = 12$ km, $w = 4$ kN/m², C_L^* and β_m as given for the Lightning in Figure 10.14.

Turning in supersonic flight

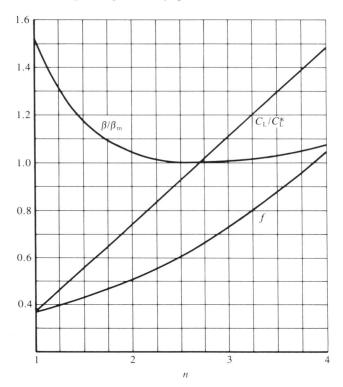

Figure 10.22. Conditions in turns represented in Figure 10.21, for $M = 1.8$.

EXAMPLE 10.4. THRUST REQUIRED FOR TURN OF SUPERSONIC COMBAT AIRCRAFT

A combat aircraft with a wing loading of 4 kN/m^2 makes a banked turn at $M = 1.8$ and a height of 6 km. The coefficients in Equation (3.6) are $K_1 = 0.025$, $K_2 = 0.4$, as in Example 10.2. If the angle of bank is $60°$, find the radius of turn and the value of F/W required to maintain height and speed.

Preliminary calculations are needed to find the values of C_L^*, β_m and C_L. These give

$$C_L^* = (K_1/K_2)^{1/2} = 0.250$$

and

$$\beta_m = 2(K_1 K_2)^{1/2} = 0.200.$$

With w measured in kN/m^2, Equation (10.1) gives

$$M^2 C_{L0} = 0.0141 \, w/\delta$$

and for $\phi = 60°$, $n = \sec\phi = 2$. Then for $h = 6$ km, for which $\delta = 0.4656$,

$$C_L = nC_{L0} = 2 \times 0.0141 \times 4/(0.4656 \times 1.8^2) = 0.0748.$$

Hence $C_L/C_L^* = 0.299$ and Equation (8.27) shows that $\beta = 1.821\beta_m = 0.3642$ and hence $F/W = f = n\beta = 0.7285$.

For $h = 6$ km the velocity of sound is

$$a = 340.3 \times 0.9299 = 316.4 \text{ m/s}.$$

Hence $V = 1.8a = 569.6$ m/s and Equation (8.7) gives

$$V^4/(g^2 R^2) = n^2 - 1 = 3.$$

Thus $R = V^2/(3^{1/2}g)$ metres $= 19.1$ km.

Note that the radius is greater than the value of 16.6 km shown in Figure 10.20 for $n = 2$ and $M = 1.8$, because at $h = 6$ km the velocity of sound is greater than it is for $h \geq 11$ km.

Appendixes

Appendix 1 List of symbols

The number given in the right-hand column shows either the section in which the symbol is first introduced or that in which further information may be found. (A2 refers to Appendix 2.)

Symbol	Description	Section
$A = b^2/S$	aspect ratio of wing	3.1.1
$A = 1 + (V/g)\,dV/dh$	acceleration factor in climb	4.6
A	propeller disc area	5.3
A	constant in equations	
a	velocity of sound in air	1.1
a_0	velocity of sound in ISA at sea level	A2
$\bar{a}$	mean acceleration during ground run in take-off or landing	6.7
b	span of wing	
C_D	drag coefficient	1.2
C_{DG}	drag coefficient during take-off ground run	6.7
C_{DL}	lift-dependent drag coefficient	3.1
C_{DLDV}	lift-dependent viscous drag coefficient	3.2
C_{DV}	vortex drag coefficient	3.1.1
C_L	lift coefficient	1.2
$C_L{}^*$	lift coefficient for minimum β	3.3
C_{L0}	lift coefficient in straight and level flight, at same speed and height as in turn	8.4
C_{L_1}	lift coefficient for minimum C_D	3.2.1
C_{LB}	lift coefficient at 'break' in C_D–$C_L{}^2$ relation for Lightning aircraft	10.1
C_{LG}	lift coefficient during take-off ground run	6.7
$C_{L\max}$	maximum lift coefficient	2.4
$(C_L)_{EF}$	lift coefficient for steady level flight at speed V_{EF}	6.7
$(C_L)_R$	lift coefficient for steady level flight at speed V_R	6.7
$(\Delta C_L)_\alpha$	increase of C_L due to flap deflection at constant α	6.1
$C_P = P_s/(\rho N^3 d^5)$	propeller power coefficient	5.3
$C_T = F_P/(\rho N^2 d^4)$	propeller thrust coefficient	5.3
$c = Q/F$	specific fuel consumption of turbojet or turbofan engine	5.1
$c' = Q/P_e$	specific fuel consumption of turboprop, related to equivalent shaft power	5.1

Symbol	Description	Section
$\bar{c}$	mean chord of wing	3.1.1
c_0	constant in Equation (5.10)	5.2.3
c_1	constant in Equation (5.11)	5.2.3
c_{CR}	sfc ($=Q/F$) at cruise rating	5.2.3
c_h	constant in Equation (5.26)	5.6
c_{so}	sfc ($=Q/F$) at take-off rating in static conditions at sea level	5.2.3
$c_s' = Q/P_s$	specific fuel consumption of turboprop, related to shaft power	5.4
c'_{sCR}	value of c_s' at rated cruise power	5.4.2
c'_{sTO}	value of c_s' at take-off rating in static conditions at sea level	5.4.2
D	drag	1.2
D_v	vortex drag	2.2
d	diameter of propeller	5.3
E	total energy of aircraft	4.5
E_t	endurance	7.9
$e = E/m$	specific energy of aircraft	4.5
$e = 1/k$	efficiency factor	3.2
$e_t = r_a/V$	specific endurance	7.9
F	total thrust of engines	1.2
F_0	maximum take-off thrust in static conditions at sea level	5.2.2
F_1	thrust at $v=1$	4.3
F_J	jet thrust of turboprop	5.4
F_m	maximum available thrust	5.2.2
F_P	propeller thrust	5.3
$F_s = F/\dot{m}$	specific thrust	5.1
$f = F/W$		2.1
$f_0 = F_0/W$		6.7
$f_1 = F_1/W$		4.3
G	braking force on wheels in landing	6.12
g	acceleration due to gravity	
$g_w = (V_m - V_w)/V_m$	factor used to correct take-off or landing distance for effect of wind	6.16.1
h	height above sea level	
$h_1 = 11$ km	height at lower boundary of region of constant temperature in ISA	1.1
$h_e = h + V^2/(2g)$	energy height	4.5
h_f	height at start of landing flare	6.11
h_s	pressure height (height in the ISA at which the air pressure would be equal to a measured pressure)	1.1
h_{sc}	height of screen for take-off or landing	6.5
$J = V/(Nd)$	propeller advance ratio	5.3
K_1, K_2	coefficients in Equation (3.6)	3.2
K_3, K_4, K_5	coefficients in Equations (3.9) and (3.10)	3.2.1
K_G	constant defined by Equation (6.9)	6.7

List of symbols

$K_{LDV} = C_{DLDV}/C_L^2$		3.2
K_{uc}	constant in Equation (6.1), giving drag of undercarriage	6.2
$k = \pi A K_2$		3.2
k_1	constant in Equation (6.7)	6.7
k_2, k_3	constants in Equation (5.8)	5.2.2
k_4	constant in Equation (5.10)	5.2.3
$k_v = \pi A C_{DV}/C_L^2$	vortex drag factor	3.1.1
L	lift	1.2
$L = -dT/dh$	temperature lapse rate in atmosphere	1.1
$L' = F \sin \theta_F$	component of thrust in direction of lift force	9.6
$M = V/a$	Mach number	1.2
M_D	drag divergence Mach number	2.4
M_P	maximum flight Mach number for propeller aircraft	5.3
M_T	Mach number corresponding to resultant velocity of propeller blade tip relative to air	5.3
m	total mass of aircraft	
m, n	exponents in power law equations	
$\dot{m}$	mass flow rate of air in power plant	5.1
m_f	mass of fuel used	7.4
N	rotational speed of engine or propeller (rev/s)	5.2.1
$N = V/V_s$		8.4
$n = L/W$	load factor	8.1
$n' = L'/W = f \sin \theta_F$		9.6
$n_1 = 1/v^2 = C_L/C_L^*$		3.3
P, Q	constants in Equation (3.15)	3.3
P_e	equivalent shaft power of turboprop	5.4
P_s	shaft power to propeller	5.3
P_{so}	shaft power to propeller at take-off rating at sea level	5.4.1
P_{sm}	shaft power to propeller at cruise rating	5.4.2
$P_{se} = V(F-D)/W$	specific excess power	4.5
p	air pressure	
p_0	pressure in ISA at sea level	A2
p_1	pressure in ISA at a height of 11 km	A2
p_t	pressure in pitot tube	1.3
Q	mass of fuel used per unit time	5.1
R	gas constant for air	A2
R	radius of curvature of flight path	8.1
R_a	distance travelled in still air	7.3
R_{ac}	cruise range in still air	7.4
R_{acr}	reference range (theoretical maximum cruise range)	7.8
R_g	range of aircraft relative to the ground	7.12
Re	Reynolds number	1.2

Appendix 1

r	number of steps in calculation of landing ground run	6.12
r	exponent of σ in § 7.5.4	
$r_a = -dR_a/dm$	specific range in still air	7.4
$r_g = -dR_g/dm$	specific ground range	7.12
S	gross wing area	1.2
s	general horizontal distance	6.7
s_a	horizontal length of airborne path in take-off or landing	6.6
s_B	distance in landing between touchdown and point where brakes become effective	6.12
s_{CR}	defined in Equation (7.29)	7.10
s'_{CR}	defined in Equation (7.39)	7.10
s_F	horizontal distance covered in landing flare	6.11
s_g	length of ground run in take-off or landing	6.6
s_R	length of ground run to accelerate from rest to rotation speed in take-off	6.6
$s_{RL} = s_g - s_R$	length of ground run in take-off between start of rotation and lift-off	6.6
T	air temperature	
T_0	temperature in ISA at sea level	A2
T_1	constant temperature in ISA at heights between 11 and 20 km	A2
T_s	temperature in ISA at specified pressure height	4.9
t	time	
t_{CR}	defined in Equation (7.33)	7.10
t'_{CR}	defined in Equation (7.43)	7.10
t_R	time for rotation phase of take-off	6.7
V	true air speed (TAS)	1.2
V^*	true air speed for minimum β in straight and level flight	4.6
$\bar{V}$	mean velocity over a specified interval	6.7
V_1	decision speed in take-off	6.5
V_2	take-off safety speed	6.5
V_3	speed at screen height in take-off	6.5
V_A	air speed at threshold in landing approach	6.11
V_B	air speed in landing ground run when brakes first become effective	6.12
$V_c = dh/dt$	rate of climb	2.5
$V_c' = dh_e/dt$	rate of increase of energy height	4.5
V_{cmax}	maximum rate of climb	4.2
V_{EF}	engine failure speed	6.6
V_{EFR}	speed at which engine failure is recognised in take-off	6.5
V_e	equivalent air speed (EAS)	1.3
V_e^+	EAS greater than V_{ec}	2.7
V_e^-	EAS less than V_{ec}	2.7

List of symbols

Symbol	Description	Section
V_e^*	EAS for minimum β in straight and level flight	2.5
V_{ec}	EAS for maximum $(F - D)$ in straight and level flight	2.5
V_{es}	EAS for $C_L = C_{L\max}$ in straight and level flight	2.5
V_F	mean speed in landing flare	6.11
$V_h = (1/V)\,dV/dh$		4.6
V_J	mean velocity of propulsive jet, relative to aircraft	5.1
V_{J1}	velocity of jet from core engine	5.2
V_{J2}	velocity of jet from by-pass duct	5.2
V_{LO}	lift-off speed in take-off	6.5
V_m	mean air speed used in correcting for effect of wind on take-off or landing	6.16.1
V_{MCa}	minimum control speed in the air after failure of one engine	6.5
V_{MCg}	minimum control speed on the ground after failure of one engine	6.5
V_{MU}	minimum unstick speed	6.5
V_R	rotation speed in take-off	6.5
V_{R1}	value of V_R for minimum take-off distance for V/STOL aircraft	9.5
V_s	true air speed at stall in straight and level flight	6.5
V_T	air speed at landing touchdown	6.11
V_w	wind velocity along flight path (headwind positive)	6.16
$v = V_e/V_e^*$	speed ratio	3.3
v_1	speed ratio v for given C_L with $L = W$ (in Chapter 9)	9.3
$v_w = V_w/V^*$	wind speed ratio	7.12
v_w'	(wind velocity in any direction)/V^* (in Figure 7.8)	7.12
$W = mg$	total weight of aircraft	
$w = W/S$	wing loading	2.1
X_G	net accelerating force in ground run at take-off or landing	6.7
Z	load on wheel, normal to runway	6.12
α	angle of incidence	1.2
α_i	induced angle of incidence at wing	2.2
$\beta = C_D/C_L$		2.1
β_m	minimum value of β	2.5
γ	angle of climb	2.1
γ_1	angle of climb calculated for given speed by assuming that $\cos\gamma = 1$	4.1
γ_A	angle of landing approach	6.11
γ_a	ratio of specific heat capacities for air	

288 Appendix 2

Symbol	Description	Section
γ_c	angle of steady climb, immediately after take-off	6.5
γ_G	gradient of runway, positive upward	6.7
$\delta = p/p_0$	relative pressure	5.2.1
δ_f	deflection angle of high-lift flap, relative to main wing	6.1
$\epsilon = f/\beta_m$		4.2
$\epsilon_1 = f_1/\beta_m$		4.3
η_o	overall efficiency of power plant	5.1
η_P	propulsive efficiency	5.1
η_{PR}	efficiency of propeller	5.3
η_{th}	thermal efficiency of turbojet or turbofan	5.1
$\theta = T/T_0$	relative temperature	5.2.1
θ	inclination of aircraft datum to horizontal	9.1
θ_F	upward inclination of thrust line to flight path	9.1
θ_{F1}	thrust inclination for minimum take-off distance, for V/STOL aircraft	9.5
λ	by-pass ratio of turbofan engine	5.2
μ	viscosity of air	1.2
μ_B	braking coefficient of friction between tyre and runway	6.12
μ_R	coefficient of rolling resistance of tyre	6.7
ρ	density of air	
ρ_0	density of air in ISA at sea level	A2
ρ_1	density of air in ISA at a height of 11 km	A2
ρ_s	density of air in ISA at specified pressure height	4.9
$\sigma = \rho/\rho_0$	relative density	1.3
ϕ	angle of bank	8.2
$\Omega = V/R$	rate of turn	8.4

Suffixes 1 and 2 – initial and final states in cruise or climb 4.7
Suffixes used in § 5.2.4
    DRY condition for maximum dry thrust
    RH   with maximum reheat
Suffixes used in § 7.10
    CL   climb
    CR   cruise
    D   descent

Appendix 2 The International Standard Atmosphere

The properties of the International Standard Atmosphere (ISA) are defined in ESDU 77021 (1986). The conditions at sea level are

Pressure $= p_0 = 1.013\,25 \times 10^5$ N/m^2
Density $= \rho_0 = 1.225$ kg/m^3
Temperature $= T_0 = 15.0°$C $= 288.15$ K
Velocity of sound $= a_0 = 340.3$ m/s.

As the equations of Chapter 1 give pressure, density and velocity of sound

The International Standard Atmosphere

in terms of temperature it is useful to state first the assumptions made for variation of temperature. From sea level ($h = 0$) to a height of 11 km the lapse rate L (positive when $dT/dh < 0$) is assumed to be constant and equal to 0.0065 K/m (see Equation (1.2)). Between heights of 11 and 20 km the temperature is constant and equal to the consequential value for 11 km, namely

$$15.0 - (6.5 \times 11) = -56.5°C = 216.65 \text{ K},$$

but the lapse rate changes sign as the height increases through 20 km and the temperature *rises* linearly from 216.65 K at 20 km to 228.65 K at 32 km.

The other constants required for calculating the properties of the ISA are given in ESDU 77021 (1986) as

$$g = 9.806\,65 \text{ m/s}^2$$

and

$$R = 287.053 \text{ J/kg K}.$$

Using these constants and the equations given in Chapter 1 the functions for p, a and ρ can be derived, but there remains a choice in expressing these functions either explicitly in terms of h or in terms of the temperature ratio which is specified by the ISA definition. The two atmospheric bands for heights up to 20 km are treated separately below, with the height h expressed in *kilometres*.

(i) $0 \leq h \leq 11$ km

The temperature function in this range of heights is given directly by Equation (1.2) as

$$T = (15.0 - 6.5h)°C = (288.15 - 6.5h) \text{ K}$$

from which the temperature ratio is found as

$$T/T_0 = 1 - h/44.331. \tag{A.2.1}$$

Equations (1.5), (1.6) and (1.9) then lead to

$$p/p_0 = (T/T_0)^{5.256} \tag{A.2.2}$$

$$\rho/\rho_0 = (T/T_0)^{4.256} \tag{A.2.3}$$

and

$$a/a_0 = (T/T_0)^{1/2}. \tag{A.2.4}$$

Each of these last three equations can be used with Equation (A.2.1) to give p, ρ or a explicitly in terms of h. The exponents 5.256 and 4.256 in Equations (A.2.2) and (A.2.3) are derived from Equations (1.5) and (1.6), using the constants given earlier.

(ii) $11 \leq h \leq 20$ km

This is the constant temperature region in which

$$T = T_1 = -56.5°C = 216.65 \text{ K},$$

giving the ratio

$$T/T_0 = 0.751\,87 \tag{A.2.5}$$

Table A.2.1. *Properties of the International Standard Atmosphere*

h (km)	T (°C)	a/a_0	p/p_0	ρ/ρ_0
0	15.0	1.0000	1.0000	1.0000
1	8.5	0.9887	0.8870	0.9075
2	2.0	0.9772	0.7846	0.8216
3	−4.5	0.9656	0.6919	0.7421
4	−11.0	0.9538	0.6083	0.6687
5	−17.5	0.9419	0.5331	0.6009
6	−24.0	0.9299	0.4656	0.5385
7	−30.5	0.9177	0.4052	0.4812
8	−37.0	0.9053	0.3513	0.4287
9	−43.5	0.8927	0.3034	0.3807
10	−50.0	0.8800	0.2609	0.3369
11	−56.5	0.8671	0.2234	0.2971
12	−56.5	0.8671	0.1908	0.2537
13	−56.5	0.8671	0.1629	0.2167
14	−56.5	0.8671	0.1392	0.1851
15	−56.5	0.8671	0.1189	0.1581
16	−56.5	0.8671	0.1015	0.1350
17	−56.5	0.8671	0.0867	0.1153
18	−56.5	0.8671	0.0741	0.0985
19	−56.5	0.8671	0.0633	0.0841
20	−56.5	0.8671	0.0540	0.0719

and this is the particular value to be used in Equations (A.2.2) and (A.2.3) when determining the base values p_1 and ρ_1 for $h = 11$ km.

Equation (1.9) shows that the speed of sound depends only on basic constants and temperature, so throughout this upper height range this speed is constant and is given by

$$a/a_0 = 0.8671, \qquad (A.2.6)$$

whereas pressure and density are still variable with height as seen, for instance, in Equation (1.8). These two quantities require base values p_1 and ρ_1 for the height of 11 km, as mentioned above. Equations (A.2.2), (A.2.3) and (A.2.5) show that these are given by

$$p_1/p_0 = 0.2234 \quad \text{and} \quad \rho_1/\rho_0 = 0.2971 \qquad (A.2.7)$$

and thus Equation (1.8) gives

$$p/p_1 = \rho/\rho_1 = \exp[-0.15769(h - 11)]. \qquad (A.2.8)$$

Table A.2.1 gives properties of the ISA calculated from these equations for heights up to 20 km. Tables extending to greater heights are given in ESDU 77021 (1986).

A simple approximate equation for the density ratio was suggested by the late Professor A. R. Collar. This is

$$\sigma = \rho/\rho_0 = (20-h)/(20+h), \quad (A.2.9)$$

where again h is in km. It is correct within 0.5% for all heights up to 9 km and within 2.5% up to 13 km. The equation may sometimes be useful for quick rough calculations, as the density ratio is often required for calculations relating to lift and drag coefficients.

Appendix 3 Conversion factors
Pressure: $1 \text{ kN/m}^2 = 20.885 \text{ lb}_f/\text{ft}^2 = 0.145\,04 \text{ lb}_f/\text{in}^2$
Density: $1 \text{ kg/m}^3 = 0.062\,428 \text{ lb}_m/\text{ft}^3 = 1.9403 \times 10^{-3} \text{ slugs/ft}^3$
Distance: $1 \text{ km} = 3280.84 \text{ ft} = 0.539\,96$ international nautical miles
Velocity: $1 \text{ m/s} = 3.2808 \text{ ft/s} = 1.9438$ knots

References

Babister, A. W. (1980). *Aircraft Dynamic Stability and Response.* Oxford: Pergamon Press.
Bass, R. M. (1982). Propeller performance prediction and design techniques. In: Propeller performance and noise, *VKI Lecture Series* 1982–08. Rhode St Genese, Belgium: Von Karman Institute for Fluid Dynamics.
Bowes, G. M. (1974). Aircraft lift and drag prediction and measurement. In: Prediction methods for aircraft aerodynamic characteristics, *AGARD Lecture Series* **67**.
Buckingham, W. R. & Lean, D. (1954). Analysis of flight measurements on the airborne path during take-off. *Aeronautical Research Council Current Paper* No. 156. London: H.M.S.O.
Callaghan, J. G. (1974). Aerodynamic prediction methods for aircraft at low speeds with mechanical high lift devices. In: Prediction methods for aircraft aerodynamic characteristics, *AGARD Lecture Series* **67**.
Clancy, L. J. (1975). *Aerodynamics.* Harlow, Essex: Longman.
Cohen, H., Rogers, G. F. C. & Saravanamuttoo, H. I. H. (1987). *Gas Turbine Theory,* 3rd edn. Harlow, Essex: Longman.
Collingbourne, J. (1970). A survey of available data on the value of rolling resistance on hard runways. *RAE Technical Memorandum Aero 1233.*
Coppi, C. N. (1980). Preliminary design studies and evolution of the Gulfstream III design. In: The Grumman Aerospace and Gulfstream American Gulfstream III case study in aircraft design, Section II, *AIAA Professional Study Series.* New York: AIAA.
Definitions Panel, Aeronautical Research Council (1958). Definitions to be used in the description and analysis of drag. *Aeronautical Research Council Current Paper* No. 369. London: H.M.S.O. Also *Journal of the Royal Aeronautical Society* **62**, 796–801.
Dekker, F. E. D. & Lean, D. (1962). Take-off and landing performance. In: *AGARD Flight Test Manual.* Vol 1, Performance. Oxford: Pergamon Press.
Dornier GmbH. (1986). *Dornier* 228 *Specification,* Appendix C, Performance. Issue 3. Munich: Dornier GmbH, Aircraft Sales.
Duncan, W. J., Thom, A. S. & Young, A. D. (1970). *Mechanics of Fluids.* London: Edward Arnold.
ESDU 68046 (1977). Atmospheric data for performance calculations. Performance Vol 1.
ESDU 71025 (1988). Frictional and retarding forces on aircraft tyres. Part I: Introduction. Performance Vol 3.
ESDU 71026 (1988). Frictional and retarding forces on aircraft tyres. Part II: Estimation of braking force. Performance Vol 3.

ESDU 72023 (1972). Low-speed longitudinal aerodynamic characteristics of aircraft in ground effect. Aerodynamics Vol 9a.
ESDU 73018 (1980). Introduction to estimation of range and endurance. Performance Vol 5.
ESDU 73019 (1982). Approximate methods for estimation of cruise range and endurance: aeroplanes with turbojet and turbofan engines. Performance Vol 5.
ESDU 75018 (1975). Estimation of cruise range: propeller-driven aircraft. Performance Vol 5.
ESDU 76034 (1976). Estimation of take-off thrust using generalised data for turbojet and turbofan engines. Performance Vol 2.
ESDU 77021 (1986). Properties of a standard atmosphere. Aerodynamics Vol 1b. See also ESDU 77022 (1986). Equations for calculation of International Standard Atmosphere and associated off-standard atmospheres. Performance Vol 1.
ESDU 79015 (1987). Undercarriage drag prediction methods. Aerodynamics Vol 6.
ESDU 81026 (1986). Representation of drag in aircraft performance calculations. Performance Vol 2.
ESDU 83001 (1983). Approximate parametric method for propeller thrust estimation. Performance Vol 2.
ESDU 85029 (1985). Calculation of ground performance in take-off and landing. Performance Vol 4.
ESDU 87037 (1988). Take-off performance of vectored-jet-thrust aircraft. Performance Vol 4.
ESDU EG 5/1 (1972). Estimation of take-off distance. Performance Vol 4.
ESDU EG 6/3 (1960). Estimation of airborne distance during landing. Performance Vol 6.
ESDU EG 6/4 (1971). Estimation of ground run during landing. Performance Vol 4.
Etkin, B. (1972). *Dynamics of Atmospheric Flight*. New York: John Wiley.
Federal Aviation Administration. *Federal Aviation Regulations* (FAR). Washington: Department of Transportation (continually updated).
Fiddes, S. P., Kirby, D. A., Woodward, D. S. & Peckham, D. H. (1985). Investigation into the effects of scale and compressibility on lift and drag in the RAE 5m pressurised low-speed wind tunnel. *Aeronautical Journal*, **89**, 93–108.
Fozard, J. W. (1986). Harri-ing Sir Isaac vertically: secrets of engineering elegance in jumping jets. *Proceedings of the Royal Institution of Great Britain*, **58**, 241–90.
Gordon, B. J. (1988). The development of the unducted fan. *Aerospace*, **15**, 22–6.
Hanke, C. R. & Nordwall, D. R. (1970). The simulation of a jumbo-jet transport aircraft. Vol 2: Modelling data. *NASA CR 114 494*.
Hill, P. G. & Peterson, C. R. (1965). *Mechanics and Thermodynamics of Propulsion*. Reading, Mass: Addison-Wesley.
Hoerner, S. F. & Borst, H. V. (1975). *Fluid-Dynamic Lift*. Brick Town, New Jersey, U.S.A.: Mrs L. A. Hoerner.
Holford, J. F., Peckham, D. H. & Cook, T. A. (1972). Estimation of lost range and time in climb and descent. *RAE Technical Memorandum* Aero 1428.
Joint Airworthiness Authority. *Joint Airworthiness Requirements* (JAR). London: Civil Aviation Authority (continually updated).
Küchemann, D. (1978). *The Aerodynamic Design of Aircraft*. Oxford: Pergamon Press.
Lan, C. E. & Roskam, J. (1981). *Airplane Aerodynamics and Performance*. Ottawa, Kansas: Roskam Aviation and Engineering Corporation.
Langfelder, H. (1974). Designing for manoeuvrability – requirements and limitations. In: Preliminary aircraft design, *AGARD Lecture Series* **65**.
Liepmann, H. W. & Roshko, A. (1957). *Elements of Gasdynamics*. New York: John Wiley.

Lighthill, M. J. (1952). On sound generated aerodynamically. I. General theory. *Proceedings of the Royal Society* A, **211**, 564–87.
Lock, R. C. (1986). The prediction of the drag of aerofoils and wings at high subsonic speeds. *Aeronautical Journal*, **90**, 207–26.
Mabey, D. G. (1973). Beyond the buffet boundary. *Aeronautical Journal*, **77**, 201–15.
Mattingly, J. D., Heiser, W. H. & Daley, D. H. (1987). *Aircraft Engine Design*. Washington: AIAA.
McIntosh, W. & Wimpress, J. K. (1975). Prediction and analysis of the low speed stall characteristics of the Boeing 747. In: Aircraft stalling and buffeting, *AGARD Lecture Series* **74**.
Neumark, S. (1957). Problems of longitudinal stability below minimum drag speed, and theory of stability under constraint. *Aeronautical Research Council Reports and Memoranda* No. 2983. London: H.M.S.O.
Pankhurst, R. C. & Holder, D. W. (1952). *Wind-Tunnel Technique*. London: Pitman.
Peckham, D. H. (1970). Range performance in cruising flight. *RAE Technical Memorandum Aero 1194*.
Perry, D. H. (1969a). An analysis of some major factors involved in normal take-off performance. *Aeronautical Research Council Current Paper* No. 1034. London: H.M.S.O.
Perry, D. H. (1969b). The airborne path during take-off for constant rate-of-pitch manoeuvres. *Aeronautical Research Council Current Paper* No. 1042. London: H.M.S.O.
Pinsker, W. J. G. (1969). The landing flare of large transport aircraft. *Aeronautical Research Council Reports and Memoranda* No. 3602. London: H.M.S.O.
Pinsker, W. J. G. (1972). Glide-path stability of an aircraft under speed constraint. *Aeronautical Research Council Reports and Memoranda* No. 3705. London: H.M.S.O.
Pinsker, W. J. G. (1975). Active control technology as an integral tool in advanced aircraft design. In: Impact of active control technology on airplane design, *AGARD Conference Proceedings* **157**.
Poisson-Quinton, Ph. (1968). Introduction to V/STOL aircraft concepts and categories. In: The aerodynamics of V/STOL aircraft, *AGARDograph* 126.
Roskam, J. (1986). *Airplane Design Part III*: Layout design of cockpit, fuselage, wing and empennage: cutaways and inboard profiles. Ottawa, Kansas: Roskam Aviation and Engineering Corporation.
Rutowski, E. S. (1954). Energy approach to the general aircraft performance problem. *Journal of the Aeronautical Sciences*, **21**, 187–95.
Saravanamuttoo, H. I. H. (1987). Modern turboprop engines. *Progress in Aerospace Sciences*, **24**, 225–48.
Schlichting, H. & Truckenbrodt, E. (1979). *Aerodynamics of the Airplane*. Translated from the German by H. J. Ramm. New York: McGraw-Hill.
Seckel, E. (1975). The landing flare: an analysis and flight-test investigation. *NASA CR-2517*.
Smith, A. M. O. (1975). High-lift aerodynamics. *Journal of Aircraft*, **12**, 501–30.
Smith, M. J. T. (1989). *Aircraft Noise*. Cambridge: University Press.
Taylor, D. R. (1975). Payload without penalty – a suggestion for improving the take-off performance of fixed-wing V/STOL aircraft. *Aeronautical Journal*, **79**, 344–8.
Thwaites, B. (Editor) (1960). *Incompressible Aerodynamics*. Oxford: Clarendon Press.
Tobin, J. R. (1945). Test flying. *Journal of the Royal Aeronautical Society*, **49**, 343–52.
Torenbeek, E. (1982). *Synthesis of Subsonic Airplane Design*. Delft: University Press.
Whitcomb, R. T. (1976). A design approach and selected wind-tunnel results at high subsonic speeds for wing-tip mounted winglets. *NASA TN D-8260*.

White, M. D. (1968). Proposed analytical model for the final stages of a landing of a transport airplane. *NASA TN D*-4438.

Whitlow, J. B. & Sievers, G. K. (1988). Return of turboprops – building the foundation. *Aerospace America,* **26,** 10, 15–20.

Williams, J. (1972). Airfield performance prediction methods for transport and combat aircraft. In: Aircraft performance – prediction methods and optimization, *AGARD Lecture Series* **56**.

Notes on references

AGARD   Advisory Group for Aerospace Research and Development, North Atlantic Treaty Organisation, 7 Rue Ancelle, 92200, Neuilly sur Seine, France.

AIAA   American Institute of Aeronautics and Astronautics, Inc. 1633 Broadway, New York, NY 10019.

ESDU   Data sheets published by ESDU International plc, 27 Corsham Street, London N1 6UA. The dates given are for the latest known amendment. The original date of issue may be earlier.

NASA   National Aeronautics and Space Administration, Washington, D.C.

RAE   Royal Aerospace Establishment (formerly Royal Aircraft Establishment), Farnborough, Hampshire, UK.

Reports quoted are not necessarily available to members of the public or to commercial organisations.

Index

acceleration factor, 62
acceleration normal to flight path, 215–18
accelerometer, 217
afterburning, see reheat
airborne part of take-off, 131–2, 141–6
   required climb gradient in second segment, 132, 157
   transition after lift-off, 141–6
air-breathing power plant, 74–5
Airbus with wing tip fence, 38 (photo)
air speed
   calibrated (CAS), 6
   equivalent (EAS), 5
   indicator (ASI), 6
   true (TAS), 5
airworthiness certification, 128–30
altimeter, 3
angle of
   attack, see angle of incidence
   bank, 217, 226–7, 236
   climb, 9, 21–2, 43, 49–50
      in a turn, 230–1
   glide, 10, 210
   incidence, 4–5
anti-g suit, 219
approximation $\cos\gamma = 1$ and consequent errors, 47–51
aspect ratio (of wing), 31
atmosphere, see International Standard Atmosphere (ISA)

balanced field length, 131, 156
bank angle, see angle of bank
Boeing 747, 121–2, 254–8, 265–8
boundary layers, laminar and turbulent, 16
braking, see landing ground run
Breguet range, 167–8
   equation, 167
      as indicator of design efficiency, 167–8

buffet boundary, 186
by-pass flow (in turbofan), 78
by-pass ratio, 78
   typical values, 80

ceiling
   definitions, 25–6
   in a turn, 235
chord, mean (of wing), 31
civil turbofans, 78–80
   Rolls-Royce RB211 series, 78–9, 83–90
   specific fuel consumption (sfc), 79
      at reduced thrust, 89–90
      variation with flight speed and height, 87–9
   variation of maximum thrust with flight speed, 82–6, 134–6
      height, 86–7
climb
   angle, see angle of climb
   at constant
      EAS, 62–5
      Mach number, 63–4
   at supersonic speeds, 268–73
   effect of acceleration, 62–5
   performance in terms of energy height, 66–9
   rate, see rate of climb
Concorde, 90, 96–7, 257–62, 269–72, 274–6
   use of reheat, 97, 270–1
conversion factors, 291
$\cos\gamma = 1$ (approximation) and consequent errors, 47–51
cruise-climb, 170–6, 182, 185–6
   thrust required, 172–7
   thrust power required, 182
cruise procedures, practical, 185–8
cruise range, see range in cruise

Index

density of air
  at sea level, 288
  relative, 5, 289–91
diversion to alternative airfield, 163–5
drag, 4
  at supersonic speeds, 253, 257, 259–65
  at transonic speeds, 253–7
  coefficient, 4
  components, 30–2
  datum, 31–3
  equations, 34–7, 39–40
    for take-off and landing, 126–7
    parabolic drag law; dual, 40; modified, 39; simple, 36, 253–4, 256
  in a turn, 228
  intake momentum, 80, 239
  lift-dependent, 31–2, 34–6
  polar, 10–12, 252
    at supersonic and transonic speeds, 252–5
  power, 22–4, 43–5
  rise at transonic speeds, 17–18, 253–7
  shock-induced separation, 253–4
  spillage, 31–2, 34–6
  viscous, 32, 34–5
  vortex, 12–13, 31–3, 253
  wave, 32–4, 252–4
drag/lift ratio β, 9–12
  at supersonic speeds, 260, 262, 266
  at transonic speeds, 255–6, 258
  in a turn, 230
  minimum value, 12, 40
    at supersonic speeds, 260–2, 266, 279
    at transonic speeds, 256, 258
  variation with
    C_L, 11–12, 40–2
    EAS, 14, 41
    M, 17–20, 258, 260, 262–3
    Re, 14–16
    speed ratio v, 41
  with vectored thrust, 240

EAP, British Aerospace, 215 (photo)
EAS, see air speed, equivalent
efficiency factor (aircraft), 39
efficiency of aircraft, 167–8
efficiency of power plant, 76–7
endurance, 162, 194–200
energy equations, 60–1
energy height, 60
engine failure
  causing increase of aircraft drag, 204–6
  causing reduction of cruising height, 205–6
  during take-off, 130–2, 137–8, 155–6
  effect on cruise range, 205–7

engine weight, reduction for small loss of range, 177

flaps, see high-lift devices
forces on aircraft in steady straight flight, 8–9
fuel allowances, 164–5
fuel used in climb between specified energy heights, 69
functional relations, turbojet or turbofan, 80–2

GE-UDF® engine, 110–11
g suit, see anti-g suit
ground proximity, effects on lift and drag, 125–6
Grumman Gulfstream with winglet, 38 (photo)

Harrier, aircraft with vectored thrust, 237–8, 245–7
Hawk, advanced training aircraft, 52–3
high-lift devices, 120–4
  effects on
    drag, 122–4
    maximum lift coefficient, 120–3
  flaps
    at leading edge, 122
    at trailing edge, 120–3
  slats, 120–3

induced incidence, 13
intake momentum drag, 80, 239
International Standard Atmosphere (ISA), 1–2, 288–91

KE factor, see acceleration factor
knots, 6, 291

landing
  airborne distance from threshold, 149–50
  approach, 147
  baulked, 154–5
  distance, effects of air density, 158–9
  discontinued approach, 154–5
  flare, 147–50
    difficulty of controlling, 148–9
  float before touchdown, 148
  ground run, 151–4
    brake torque limit, 153
    braking friction coefficient, 152–3
    braking parachutes, 151–2

Index

landing ground run—(*Contd.*)
  mean acceleration, 153
  reversed thrust, 151, 154
  rotation after touchdown, 151
landing—(*Contd.*)
  threshold, 147, 150
  vertical, 245
lift, 4
  coefficient, 4
    for minimum β, 40; at supersonic speeds, 260–2, 266
    in steady straight flight, 10
    maximum, 11–12, 120–3, 125; variation with Mach number, 19–20, 219–20, 277
lifting-line theory, 31, 33
lift-off speed in take-off, 128–9
lift/weight ratio, *see* load factor in a turn
  with vectored thrust, 240
Lightning, English Electric supersonic fighter, 257–8, 261–6, 280–1
load factor in a turn, 215, 218
  human limitations, 219
  stalling and buffeting limits, 219–23, 226–7
    at supersonic speeds, 277–9
    dependence on Mach number, 219–23
  structural limitations, 218
  thrust limitations, 233–5
    at supersonic speeds, 279–82
Lockheed SR-71 Blackbird, 25–6
loop manoeuvre, 216
lost fuel in climb and descent, 200–4
lost range and lost time in climb and descent 163, 200–4

Mach number, 5–6, 17
  drag divergence, 17–18
military turbofans, 90–6
  variation of maximum thrust with flight speed, 91–5, 135
  specific fuel consumption (sfc), at reduced thrust, 93, 95–6
  variation with flight speed, 91–5
minimum control speed, in flight and on the ground, 128–9
minimum drag, *see* drag/lift ratio β, minimum value
minimum unstick speed, 128–9
minimum work for given range, 164–6

nautical miles, 6, 291
noise reduction due to by-pass flow, 79
non-dimensional relations, turbojet or turbofan, 80–2
normal acceleration in curved flight, 215–18

Olympus 593 turbojet, 96–7
open rotors, *see* propfans
Osprey, Bell Textron tilt-rotor aircraft, 237–8
overall efficiency of power plant, 76–7

partially jet-borne (PJB) flight, 242–4
payload–range diagram, 212–13
phases of flight using fuel, 163–5
piston engines, 115–16
  specific fuel consumption (sfc), 116
  variation of power with height, 116
pitot tube, pressure in, 6
pressure height, 3
propellers, 97–101
  advance ratio, 97
  contra-rotating, 99, 101
  disc loading, 98–9, 101
  efficiency, 97–101
    variation with advance ratio, 98–100
  interference with aircraft, 99–100
  Mach number, blade tip, 98
    effect on; efficiency, 98, 100–1; noise, 101
  power and thrust coefficients, 97
  upper limit of flight Mach number, 101, 185–7
  variable blade angle, advantages, 99
propfans, 108–15
  contra-rotating, 110–11
  specific fuel consumption (sfc), 114–15
    at reduced thrust, 115
    variation with flight speed and height, 114
  variation of maximum thrust with flight speed and height, 112–13
propulsive efficiency, 76
propulsive system, 75–6
pull-out from a dive, 215–17

range, 162
  effects of climb and descent, 200–3
  effect of varying payload, 212–13
  in cruise, 166–9
    calculation of, 188–94, 276–7
  conditions for maximum
    at constant height, 177–85
    at constant Mach number, 172
    at constant speed, 170–1
    at constant thrust power, 184–5
    at supersonic speeds, 273–7
    at transonic speeds, 263, 265–8
    when limited by available thrust, 173–7
rate of climb, 22–4, 42–3, 49–59
  at supersonic speeds, 268–73

Index

correction for non-standard atmosphere, 72–3
effect of acceleration, 62, 64–5
in a turn, 228–31
rate of turn, 224–7, 235–6
reference speeds for take-off, 128–30
reheat (afterburning), 80, 90, 93–5, 97
effects on maximum thrust and sfc, 93–5, 97
reserve fuel, 163–5
Reynolds number, 5
rolling resistance, 133
Rolls-Royce RB211 turbofans, 78–9, 83–90
Rolls-Royce Tay 650 turbofan, 83–4, 86, 90
rotation speed in take-off, 127, 129

SAAB 2000, turboprop aircraft, 187
screen height for take-off, 128, 131–2
sfc, *see* specific fuel consumption
simple parabolic drag law, *see* drag equations
ski-jump, for take-off with vectored thrust, 245–6
slats, *see* high-lift devices
slugs/ft^3, 291
specific endurance, 194
specific energy, 60
specific excess power (SEP), 61
in a turn, 228–33
specific fuel consumption (sfc)
turbojets and turbofans, 77, 81–2, 87–97
turboprops, 77, 102–3, 106–9
specific range, 166
specific thrust, 76
speed for
maximum angle of climb, 21–2, 69–71
effect of acceleration, 71–2
maximum endurance,
turbofans, 195–6
turboprops, 197–9
maximum rate of climb, 22–4, 51–9
effect of acceleration, 65–6
with thrust independent of speed, 51–6
with thrust power independent of speed, 56–8
with thrust power increasing with speed, 58–9
maximum specific excess power, 67–9
minimum drag power, 24, 43–5
minimum rate of sink in glide, 44
occurrence of engine failure in take-off, 137–8, 155

recognition of engine failure in take-off, 129–31, 155
speed ratio v, 41
speed stability, 27–9, 270
stalling, 11–12
speed, 120, 128
stepped cruise, 178, 186, 189, 191–2, 276

take-off
decision speed, 129–31
delay period after engine failure, 155
distance, effects of air temperature and pressure, 158
distance from rotation to lift-off, 138–40
ground run, up to rotation speed, 133–40
mean acceleration, 134, 139–40
optimum lift coefficient, 137
with engine failure, 137–8
safety speed, 128–9
short, with vectored thrust, 245–8
vertical, 244–5
thermal efficiency, turbojet or turbofan, 76–7
thrust
derived from momentum change, 75–6
for take-off, effects of air temperature and pressure, 157
gross, 80
inclination (*see also* thrust, vectored)
optimum for cruise and climb, 241
optimum for short take-off, 247
line of action, 8–9
maximum available, 74
net, 80
rating, 74
required
reduction for small loss of range, 177
to maintain height and speed in a turn, 233–6, 249–51, 279–82
rotation speed, for take-off with vectored thrust, 245–7
vectored, 237, 239–40
use in a turn, 248–50
use in combat, 251
thrust/weight ratio f, 9–10
tilt rotor and tilt wing, 237–8
time required and minimum time for climb, 24–5
climb between specified energy heights, 68–9
transonic speed range, 252–3
trimmed state of aircraft, 4
turbofans (*see also* civil turbofans and military turbofans), 78–9
mixed-stream, 78–9
turbojets, 77–8

turboprops, 102–3
  control by varying blade pitch, 103
  equivalent shaft power, 102–3
  maximum shaft power, 103–6
    variation with; flight speed, 103–4; height, 104–6
  ratio of equivalent shaft power to shaft power, 105–6
  specific fuel consumption (sfc), 106–9
    at reduced power, 107–9
    variation with flight speed and height, 106–8
turn, true-banked, 217–18
  acceleration normal to flight path, 217–18
  increase of drag, 228
  radius of curvature of flight path, 217–18, 221–4, 226–7, 236
    limitations (*see also* load factor in a turn)
    with vectored thrust, 248–9

undercarriage drag, 124–5
unducted fan (UDF) engine, 110–11

vectored thrust, *see* thrust, vectored
velocity of sound, 2
viscosity, variation with temperature, 14
vortex drag factor, 33, 36

WAT curves and WAT limit, 157
wind, effect on
  take-off distance, 159–60
  landing distance, 160–1
  speed for minimum angle of glide relative to ground, 210
  speed for maximum ground range, 208–11
wing area, gross, 4
wing loading, 10
winglets and wing fences, 37–9